An Introduction to the Physics of Semiconductor Devices

David J. Roulston

New York • Oxford
Oxford University Press
1999

Oxford University Press

Oxford New York
Athens Auckland Bangkok Bogotá Buenos Aires Calcutta
Cape Town Chennai Dar es Salaam Delhi Florence Hong Kong Istanbul
Karachi Kuala Lumpur Madrid Melbourne Mexico City Mumbai
Nairobi Paris São Paulo Singapore Taipei Tokyo Toronto Warsaw

and associated companies in
Berlin Ibadan

Copyright © 1999 by Oxford University Press, Inc.

Published by Oxford University Press, Inc.
198 Madison Avenue, New York, New York 10016
http://www.oup-usa.org

Oxford is a registered trademark of Oxford University Press

Library of Congress Cataloging-in-Publication Data

Roulston, David J.
    An introduction to the physics of semiconductor devices / David J. Roulston.
        p. cm. — (Oxford series in electrical and computer engineering)
    Includes bibliographical references and index.
    ISBN 0–19–511477–9 (cloth)
    1. Semiconductors. I. Title. II. Series.
QC611.R86   1998
621.3815'2—dc21   97–52396
                                                                   CIP

9 8 7 6 5 4 3 2 1

Printed in the United States of America
on acid-free paper

To my two families:

the Roulstons: *Christine, Hélène, Philippe*

and the Harpers: *Peter, Euan, Robin*

Contents

Chapter 3 *P–N* Junction Diodes   59

Chapter 4 Bipolar Junction Transistors   93

Chapter 5 MOS Field-Effect Transistors   137

Chapter 6 Junction Field-Effect Transistors   167

Chapter 7 Overview of Special-Purpose Semiconductor Devices   183

Chapter 8 Silicon Chip Technology and Fabrication Techniques   215

Appendixes   **241**

Preface

Background

This text is aimed at those students with no previous exposure to semiconductor devices. It is therefore primarily for second-year students of an electrical engineering, electronics, or physics curriculum. The book results from the author teaching an undergraduate course on semiconductor devices for more than 20 years at the University of Waterloo. During that time the course was sometimes preceded by a "Modern Physics" course, sometimes by a "Circuits" course, depending on the way the overall curriculum evolved over the years. These factors have led to the present work being produced as a self-contained text that does not require any circuits background, nor any modern physics background. Only a first-year physics course and a course in calculus including differential equations are assumed as prerequisites, as is an introductory course on basic electrical engineering topics. Some elementary circuits background is given in the first chapter so that students will know how the devices are to be used. A summary of quantum physics results is included in the first six appendixes, but this need only be consulted for background information and is not considered part of a course on "devices." If the students have already had exposure to modern physics and/or to circuits, the first six appendixes and much of Chapter 1 may be omitted.

A guiding philosophy in writing the text has also been that if the mathematics is kept simple, it is easier for the student to grasp the underlying physical concepts. Because of the author's involvement with industry, through its use of the device simulation program *BIPOLE*, it also became apparent that some of the mathematical treatment in existing texts introduced needless complications that detract from understanding the basic principles. An example is the widespread inclusion in many existing texts of recombination (leading to unwieldy and confusing hyperbolic functions) when formulating current flowing to the base of a bipolar transistor. In industrial devices this is usually a negligible effect, and its exclusion allows simpler equations to be used and hence more attention to be paid to more important effects, like band gap narrowing and recombination in the emitter.

At the same time, a subsidiary goal was to keep the text concise and (hopefully) economical. This is consistent with the above-mentioned objective of keeping the material clear and the mathematics simple.

Organization of Material

As mentioned, Chapter 1 gives a survey of the basic circuit elements: diodes and transistors, with their associated simple single device circuit at the most elementary level. This chapter also provides an overview of semiconductor material and device properties at an elementary level. Although this involves some slight repetition in later chapters, it provides a useful summary for students to refer to and so to keep in mind the overall objectives of the course. These objectives can otherwise often become obscured when dealing with details of the physics of specific devices.

The physics background material of Chapter 1 is expanded in Chapter 2 to cover material properties and some general aspects of *PN* junctions. The concepts of band gap, intrinsic carrier concentration, and doping are treated without recourse to quantum mechanics. This is explained on the basis of *models* whose characteristics may be obtained by electrical measurements. The use of this approach has been based on teaching such a course for over 20 years, where it has been found that attempting to carry out quantum mechanical derivations for second-year students often leaves them confused with the physics and mathematics, instead of enabling them to focus on the electrical characteristics of semiconductor devices. As mentioned, a summary of the essential quantum mechanical treatments is included in Appendixes A through F, which may be used if the instructor deems this appropriate for the course.

Chapter 3 covers various types of diode, chosen for both pedagogical and practical reasons; for example, the narrow-base diode is highlighted, because (1) the treatment is exceedingly simple, and (2) it forms the basis of the treatment for the bipolar transistor. Both dc, small-signal, and transient characteristics of diodes are covered. Specific diode types studied include the narrow-base and *PIN* diodes, with only a brief reference to the wide base and P^+NN^- structures. The metal–semiconductor contact and the Schottky diode are also included.

In Chapter 4 the bipolar transistor is examined. Using realistic approximations valid for most industrial BJTs, the mathematics is kept to a strict minimum without any loss of usefulness (as stated); nevertheless all important aspects of BJT characteristics and operation are introduced including dc, small-signal ac, and transient operation. The important figures of merit f_t and f_{max} are explained, and a short overview of noise performance is given (with a detailed overview in Appendix N).

Chapter 5 is devoted to MOSFET devices. A study of the MOS system and the MOS capacitance is presented, followed by a conventional treatment using the Shockley theory. After deriving the dc and small-signal characteristics, second-order effects, especially those related to short-channel devices, are discussed in sufficient detail for an introductory course. Velocity saturation in the channel is highlighted to explain the nature of "pinchoff" and to distinguish between long- and short-channel effects. The general relations among transconductance, input capacitance, and transit time are explained. The unity current gain frequency f_t is discussed, as is the switching behavior of MOS and CMOS logic inverters.

Chapter 6 covers the JFET. This chapter is shorter than that covering MOSFETs, both because the JFET is essentially a simpler device to understand and also because it is less important for the general student of electronics. The treatment explains the dc characteristics qualitatively and then covers the conventional gradual channel approximation theory. The general relationships among transconductance, channel transit time, and input capacitance is derived, and special attention is paid to the nature of pinchoff in the "short-channel" device.

Chapter 7 introduces the reader to various "special-purpose" semiconductor devices, mainly from a qualitative viewpoint. This starts with optoelectronic devices, including the photodiode and solar cell, the light-emitting diode, and the laser diode; a brief introduction is also given to charge-coupled devices (CCDs) for imaging arrays. Sections on high-power devices follow; these include the thyristor or SCR and high-power MOS devices. The insulated gate bipolar transistor is introduced. Various microwave devices are covered briefly, including the heterojunction bipolar transistor. The chapter ends with a short overview of EPROM memory devices.

It is important for students studying the physics of semiconductor devices to know how the device is made and to be aware of fabrication technology and its limitations. Chapter 8 is included for this purpose. It provides a short introduction to fabrication technology at a level suitable for a first text on devices. The basic laws of impurity atom diffusion and ion implantation are introduced; oxide growth is briefly explained; masking and lithography are summarized. Examples are given of conventional and advanced BJT and MOS structures.

A comprehensive set of Appendixes covers in some detail material that is not germane to a first course on semiconductor devices, but that the advanced (or curious) student may consult for further background and that certain instructors may wish to include as part of the formal lecture material. Appendixes A–F cover the basic quantum physics of semiconductor devices. The remaining appendixes cover theoretical aspects of specific semiconductor devices. The decision to place a considerable amount of material in the appendixes was made in order to present a smooth flow of the more basic material for the student who is studying semiconductor devices for the first time. The text contains many worked examples, and important equations are highlighted.

According to the time available, instructors may wish to shorten the material covered. This may be done, for example, by omitting the following sections: 2.4.6 (linearly graded junction), 3.3 (*PIN* diodes), 3.5.1 (wide-base diodes), 3.5.2 (P^+NN^+ diodes), 3.9 (diode switching), 3.11 (metal–semiconductor contacts), 4.5 (high- and low-current effects in the BJT), 4.8 (BJT switching), 4.11 (BJT noise performance), parts of 5.2 (the MOS system), 5.7 (second-order effects in MOSFETs), a substantial part of Chapter 6 (the JFET), any or all parts of Chapter 7, and some parts of Chapter 8. On the other hand, if more time is available, or if the level of the students is sufficiently advanced, extensive use may be made of the material in the appendixes to cover in more depth some of the background theory.

Software BIPOLE Package Accompanying This Text as CD-ROM

The Student BIPOLE plus MOSFET Option software is included with this book as a CD-ROM. It is for PC use and the graphics post-processor BIPGRAPH is supplied in both DOS and Windows 95 versions. The software contains many examples of diode, BJT, and MOSFET devices which have been simulated with the Student BIPOLE (with MOS option) or BIPOLE3 programs. For each device, a large number of graphs may be examined; these provide both internal and terminal characteristics. The given device input files may then be altered and Student BIPOLE executed to study the effects of parameter changes on device internal and terminal characteristics.

For the diode, graphs include: impurity profile, free-carrier concentration and electric field distributions, depletion-layer characteristics, and *J–V* characteristics.

For the BJT, the plots are available at different bias levels, plus many graphs to illustrate essential vertical and horizontal current flow properties. Base resistance variation with bias is included, as are terminal characteristics including I_C and I_B Gummel plots, f_t, f_{maxosc}, and current gain plots.

For the MOSFET, plots include surface electric field and carrier velocity distributions, I_{DS}-V_{DS} and I_{DS}-V_{GS} characteristics for both long-channel and short-channel devices. Also included are detailed plots of the MOS capacitance (showing the effects of surface charge and channel doping), inversion-layer charge, and surface potential versus gate bias.

Ths Student BIPOLE Examples documentation is included to enable the student to examine the detailed operation of each device as it pertains to the theory presented in the text. Appendix Q provides details about the CD-ROM.

References to the examples supplied on the CD-ROM are made at various parts of the text so the student may examine characteristics in more detail for real devices, superimpose plots from several devices, and alter the input files and re-run Student BIPOLE to study the effects of changing device parameters or bias conditions.

Users of the Student BIPOLE and BIPOLE3 software are encouraged to check periodically the BIPSIM web site at WWW.BIPSIM.COM for further documentation and information on updates.

Acknowledgments

The author has benefited from many people during the preparation of this text. In particular he would like to thank Ryan Ferguson, who not only participated as an extremely competent teaching assistant for the course on which this material was derived, but who did a splendid job of critically reading the typescript and contributing invaluable comments and suggestions. He also made considerable contributions to the development of the current version of Student BIPOLE. Thanks are also due to Mani Vaidyanathan and Andrew Sarangan who, as teaching assistants, were of great assistance in providing feedback about the problems and solutions in the early stages of this work. I also wish to thank the many other teaching assistants involved with the course from which this text evolved. It must also be said that without the enthusiastic participation of students over the past 20 years this text would not have been written.

The author has also benefited considerably over the years from discussions with his many university colleagues at Waterloo and Oxford. He thanks particularly Savvas Chamberlain, Arokia Nathan, and C. R. Selvakumar at Waterloo, and Roger Booker at Oxford. During consulting activities with SGS Thomson Microelectronics (Grenoble and Crolles, France), much has been learned from working in close collaboration over many years with Maurice Depey and Didier Celi, to whom grateful acknowledgment is made. Thanks also to my son Philippe for very substantial contributions to developing figures and documentation. Finally, thanks are due to the staff at Oxford University Press, New York for their encouragement and patience in the development of this book.

The graphs in Figures 2.8(b), 2.11, 2.12, 4.8, 4.20, 4.26, 4.33, 5.4, 5.9, 5.16(a), 5.16(b), and 7.4 were generated using GNUPLOT (copyright © 1986–1993 Thomas Williams, Colin Kelley).

Oxford, England   D. J. R.
June 1998

An Introduction to the Physics of Semiconductor Devices

Chapter 1

Overview

1.1 INTRODUCTORY REMARKS

In this chapter we give the reader a summary overview of basic semiconductor properties and devices. This material will be covered in more detail in subsequent chapters. This overview is important, since the subject matter of semiconductor devices is complex, and it is easy to lose sight of the overall picture, including learning goals, when immersed in the intricate details of "*P–N* junction injection law," "Poisson's equation using the depletion approximation," "velocity saturation in the MOSFET," etc. If the reader does not retain an awareness of this "overall picture," it is easy to become confused with the (albeit necessary) details, including a rather large number of symbols and numerous subscripts, each of which has a very specific meaning, intended to help, not hinder, in understanding the material. So the reader is strongly urged to read this chapter, and even to refer back to it frequently when it is not clear in what direction a particular section appears to be heading.

In this and following chapters, reference will be made to the appendixes. These contain background material, derivations, etc., which are not essential to understanding and using the results presented in the main body of the text but may be used optionally for a more thorough understanding in some places.

1.2 HISTORY OF SEMICONDUCTOR DEVICES

The first semiconductor device was probably the crystal detector used in the radio "crystal set" receiver in the first decades of the twentieth century. This consisted usually of a selenium crystal and a fine metal contact, adjusted manually for the best detected signal. In the 1930s and 1940s the need for a good detector diode at radar frequencies gave a huge impetus to semiconductor device research and development. In 1948 at Bell Laboratories, Murray Hill, New Jersey, Shockley, Brattain, and Bardeen invented the bipolar transistor, for which they subsequently received the Nobel Prize. The first silicon integrated circuit was invented simultaneously by Kilby at Texas Instruments and by Noyce at Fairchild in 1959. It is probably in this integrated circuit or silicon chip form that semiconductor devices are most widely known today; it should not be forgotten, however, that a large number of so-called "discrete" transistors and diodes exist for applications from microwave diode attenuators and detectors to high-voltage (thousands of volts), high-current (hundreds of amps) switching devices.

The understanding of the theoretical aspects of semiconductor material and devices depended upon the evolution of quantum mechanics. This could be said to have started with Planck's black body radiation law in 1900, Einstein's photoelectric effect in 1906 (making possible the direct measurement of Planck's constant h), followed by by the work of Rutherford and Bohr on the atom. The essential quantum mechanical theory followed rapidly with de Broglie's wave–particle duality concept (1924), Schrödinger's wave equation (1926), Heisenberg's uncertainty principle (1927), Pauli's exclusion principle, and the work of Fermi and Dirac.

1.3 WHAT IS A SEMICONDUCTOR?

1.3.1 Energy bands

In order to study the properties of semiconductor devices (diodes, transistors, integrated circuits), it is essential to possess some knowledge of the materials used in making these devices. The main material used today is silicon (in spite of the remarkable progress made in recent years in gallium arsenide and other technologies). What is this element and why is it so widely used? The simple answer is that silicon is a column IV element of the periodic table and is a semiconductor. The resistivity of semiconductors may be controlled by the addition of impurities; this also allows *pn* junctions to be formed. Silicon is the most widely used semiconductor because of the technology, which lends itself to making large numbers of devices on a wafer at economic cost.

In order to understand the nature of semiconductor material we must go back to some basic physics. Let us recall first that at the beginning of the twentieth century, the Danish physicist Niels Bohr, in attempting to explain the observed experimental results of line spectra, proposed a model for the hydrogen atom. In this model, he postulated that the orbit of the electron around the nucleus of the H atom could only lie in one of a fixed number of "orbitals" of given radii r_1, r_2, r_3, etc. Each orbital corresponds to a particular energy level, E_1, E_2, E_3. For the hydrogen atom his result for energy and the corresponding radius for the nth orbit were given as (see Appendix A for the derivation):

$$E_n = m_0 q^4 / (8\epsilon_0^2 h^2 n^2) \tag{1.1}$$

$$r_n = \epsilon_0 n^2 h^2 / (q^2 \pi m_0) \tag{1.2}$$

where ϵ_0 is the permittivity of free space, q is the electronic charge (1.6×10^{-19} Coulombs), m_0 is the mass of the electron (9.11×10^{-31} kg), and h is Planck's constant (6.63×10^{-34} J s).

The meaning of the energy of an electron in a particular orbital can best be thought of as the energy that must be given to the electron in order to separate it from the atom, that is, in order to "free" it from the bond due to the force of coulomb attraction. Figure 1.1 illustrates the main aspects of this Bohr model. Although the actual behavior of the electron orbiting the atom is much more complicated than initially proposed by Bohr, the major scientific observations, including the main features of photoabsorption and emission spectra (the "pictures" of peaks in absorption when light of varying wavelengths is shone on hydrogen gas, or the corresponding

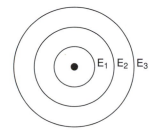

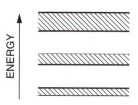

Figure 1.1 Bohr model of the hydrogen atom. E_1, E_2, E_3 are energy levels at radii r_1, r_2, r_3.

Figure 1.2 Energy bands in a solid.

peaks in emission when light is emitted (by applying some external energy source to "excite" the electrons), are explained by this very simple model.

We will be concerned in this work not with individual atoms but with groups of atoms forming the solid semiconductor. In a crystalline solid the energy orbital picture of the Bohr model of the hydrogen atom is modified in one important respect: Instead of having a series of discrete energy *levels* (as in the H atom), a crystalline solid is characterized by a series of energy *bands*, each band separated form the next one by an energy gap.

This is illustrated in Fig. 1.2. An overview of the essential quantum mechanical treatment starting with Schrödinger's wave equation is given in Appendixes B–D.

The energy diagram still has the same basic meaning as for a single atom: The value of the energy at a particular level is that energy that must be applied to the electrons at that level in order to make them escape from the solid. Perhaps the best-known engineering example of this is the vacuum tube diode (or valve), shown in Fig. 1.3, where one electrode, the cathode, is heated, thus giving the electrons in the metal enough energy to escape from the surface; by applying a positive voltage to a second electrode, the anode, the cloud of electrons having escaped from the cathode can be removed, thus producing an electron current in an external circuit.

The Einstein photoelectric experiment is a further demonstration of the existence of energy that must be applied to remove an electron from the surface of a metal. Since this experiment introduces an important relation to be used subsequently in this text, let us have a brief look at this case, where light is used as the source of energy to liberate the electrons. Figure 1.4 shows a vacuum tube with two electrodes, with a voltage applied between the cathode and anode as shown. Light of various wavelengths λ or frequencies ν ($\lambda\nu = c =$ speed of light $= 3 \times 10^8$ m/s) is incident on the cathode. As the bias potential is varied, the current flowing from anode to cathode can be brought to zero at a bias V_0. The resulting curves of current versus bias, and hence of bias V_0 versus light frequency ν, are obtained. The slope of the plot in Fig. 1.4(c) is h/q, where q is the electronic charge (1.6×10^{-19} Coulombs) and where h is

Figure 1.3 Vacuum tube. F is the heated filament, used as a cathode, from which electrons are emitted; A is the anode connected to a positive voltage.

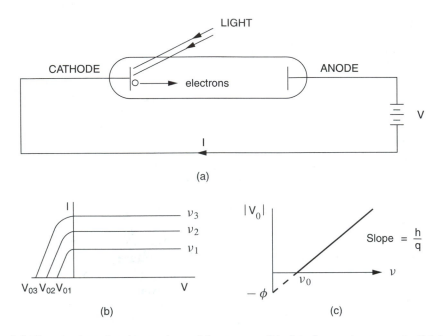

Figure 1.4 Einstein photoelectric experiment (a) apparatus; (b) plot of current versus potential V at various light frequencies ν; (c) plot of cut-off potential V_0 versus light frequency ν.

Planck's constant, which may thus be obtained. The intercept on the y axis is determined by the work function ϕ_0 of the metal. The energy of the photon as a function of frequency is given by:

$$E = h\nu \tag{1.3}$$

It is only when the energy of the incident light $h\nu_0$ is greater than the work function of the metal that electrons are liberated and current flows.

In most semiconductor devices, the most important quantity in the energy band diagram is not the energy required to *remove* the electrons completely from a certain level, but rather it is the energy required to get the electrons to move from an inner energy band to the outermost partially filled energy band; that is, it is the energy *band gap*, universally referred to by the symbol E_g, which is the single most important quantity governing the basic properties of a semiconductor; this is illustrated in Fig. 1.5.

These properties include: room-temperature conductivity (important in nearly all semiconductor devices) and photoresponse cut-off wavelength (very important in fiber optic

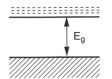

Figure 1.5 Outer two energy bands in a semiconductor showing the energy gap E_g between the conduction (upper-level) and valence (lower-level) bands.

photodetector diodes). The band gap of the most commonly used semiconductors is of order 1 electron volt (eV); for silicon $E_g = 1.1$ eV.

What do we mean by "a partially filled band"? First of all, an energy band is actually made up of a large number of discrete "energy states." Let us think back to the Bohr model of the hydrogen atom. In its normal "ground-state" condition, the H atom has just one electron occupying the innermost orbit (the properties of which we will actually use to calculate some semiconductor behavior; so let us give the values now: the ground-state energy E_1 has a magnitude of 13.6 eV and the corresponding radius r_1 is 0.53 Å or 0.053 nm). All the other orbits are empty (there are no electrons there). The inner orbit can have a maximum of two electrons, as in the helium atom (which also has a different charge in the nucleus). For atoms with more than two electrons the electrons occupy higher energy levels. In the hydrogen atom, all these outer energy levels are empty (i.e., not occupied by electrons).

In a solid we have a very similar situation. All the inner energy bands are filled (at room temperature), but all the outer energy bands are empty. As we progress from inner to outer bands, we reach a particular set called the *valence* band (almost completely filled in a pure semiconductor) and the *conduction* band (almost empty in a pure semiconductor). The conduction band is normally the one in which the electrons contribute to conduction of electric current. This occurs by electrons moving to empty energy states in response to an electric field. No current can flow if all the states in a band are full. However, we shall see that in many semiconductor devices, it is sometimes even more important to study the currrent due to the electrons in the almost filled valence band (or, as we shall see, to represent this movement of a very large number of negatively charged electrons by an equivalent concentration of positively charged particles called *holes*).

If the band gap E_g is very high (in carbon it is about 5 eV), very few electrons will be in the conduction band because the thermal energy of any particle at room temperature is of order kT, or about 0.025 eV, and very few electrons will be excited across the band gap; carbon in its crystalline diamond form is in many respects almost an insulator. However, in silicon, there will be many more electrons in the conduction band because its band gap is considerably less at 1.1 eV (as we shall see in Chapter 2, the free-electron concentration varies exponentially with $E_g/2kT$). In an insulator there are no electrons in the outer band and the next inner band is completely filled. In a metal, on the other hand, the outer (conduction) band is partially filled at all temperatures, thereby allowing conduction by the free electrons.

1.3.2 Crystals and covalent bonds

Most semiconductor devices require the use of semiconductor crystalline material as a starting point. In fact, all our discussions of energy bands and values given for electrical properties will assume that we are talking about crystalline material. What is a semiconductor crystal? Well, let us look at column IV of the periodic table, where we see, in ascending order of atomic number, the elements C, Si, Ge. These elements are represented schematically in Fig. 1.6, where each major quantum number n is shown with a corresponding number of boxes. Each box represents a quantum state with a maximum of two electrons, each with opposite spin. We need not concern ourselves here with the quantum numbers l, m_l, which correspond to small energy differences compared to the large energy differences between successive values of n. Each of the column IV elements has the property of having exactly four electrons in the outer orbit, or outer shell. It is a well-established fact (the mathematical/physics proof of which comes from quantum mechanics) that a "complete stable" outer shell contains 8 electrons.

When a solid such as silicon is formed, the individual atoms bond (i.e., are "tied together") by each valence electron being "shared" with one of four neighboring atoms in the solid, as shown in a simplified two-dimensional representation in Fig. 1.7. This is called the *covalent* bond structure. It is a very stable condition and creates a situation akin to the completely filled orbit of an atom (for example, the helium atom has two electrons in the ground-state orbit, and the energy required to remove one electron is about 24 eV). The more tightly the atoms are bound together, the more difficult it is to break the bonds and the harder is the mechanical strength of the solid. Thus carbon in its crystalline diamond form is a much harder material than silicon (which is why a diamond cutting saw can be used to slice silicon wafers). For the same reason, more thermal energy is required to move electrons from the valence band into the conduction band in carbon than for silicon. At room temperature, therefore, carbon has a much higher resistivity (much fewer electrons available for conduction) than silicon. There is a fairly direct qualitative relationship between the bond strength and the band gap. The tighter the covalent bond, the higher the band-gap energy E_g. (See Table 1.1).

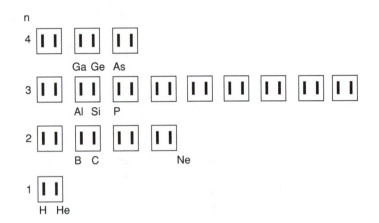

Figure 1.6 Box representation of some important elements; n is the major quantum number.

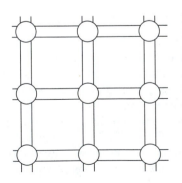

Figure 1.7 Simplified representation of the covalent bond structure in which each atom shares one of its outer electrons with each of four neighboring atoms. Each circle represents an atom; each connecting line represents a single shared electron bond.

TABLE 1.1
Band gaps of common semiconductors

Material	C	Si	Ge	GaAs	GaP	InP	InSb
Band gap at 300 K	5.47	1.12	0.66	1.42	2.26	1.35	0.17

Note: It is well worth noting that although quantum mechanics theory can be used to explain the qualitative behavior of atoms, molecules, and solids in an extremely satisfying manner, the equations become so complex (multidimensional with many variables) that it is only for relatively simple cases that quantitative results have been obtained, such as simple atoms and molecular combinations. For solids in general, quantitative results depend on the experimental determination of certain "model" parameters, one of which includes the energy band gap. This can be *measured* in a number of ways:

1. By shining light of varying wavelengths on the material and observing (e.g., from electrical conductivity measurements) the wavelength λ at which photon absorption starts to occur (as electrons are excited across the band gap by the energy $E = h\nu = hc/\lambda$);
2. By forward biasing certain *PN* junctions and observing the wavelength λ of the light emitted (as electrons drop down or "recombine" across the band gap);
3. By measuring the change in material conductivity with temperature (very easy to do in any electronics laboratory; this is discussed in Chapter 2).

Since the band gap can be easily identified from measurements, it is not essential to go into the quantum mechanics theory in great detail. The "model" is verifiable for all ranges of practical use by carrying out appropriate laboratory experiments.

We nevertheless provide an elementary treatment of the most important parts of the quantum mechanical theory in Appendix F because this will provide us with some very useful quantitative equations, as well as letting us have a clearer insight into semiconductor properties.

Electrons and holes

As mentioned, instead of considering the (small) movement of a large number of electrons in the (almost filled) valence band, it sometimes makes life a lot easier to consider the (relatively large) movement of a (smaller) number of equivalent particles called *holes*. In most semiconductor devices we shall deal with both electron and hole motion. We shall adopt the common nomenclature of calling the electron concentration n and the hole concentration p, both expressed as the number of particles per cubic centimeter. In general there will be two sets of parameter values, one for each type of particle (also referred to as current *carriers*). Thus we will talk about the values of electron and hole mobility, and electron and hole diffusion coefficient. To avoid tedious repetition we will in most cases confine our discussions to only one type of carrier, with the understanding that all remarks also apply to the other type of carrier, with appropriate values for the physical constants.

1.3.3 Doping a semiconductor with impurity atoms

All semiconductor devices depend on the fact that the (free) electron and hole concentrations can be controlled during fabrication by a process called *doping*. If we consider an element from column V of the periodic table, with 5 (valence) electrons in the outer shell, it is quite easy to accept that such an element (the most widely used are phosphorus and arsenic) can fit into a silicon crystal structure providing it gives up its extra fifth electron. This fifth electron is then *free* to conduct current. This is exactly what happens, as shown in Fig. 1.8.

Silicon can be doped with phosphorus (or arsenic) to any required impurity concentration N_D (measured in atoms per cubic centimeter), where the subscript D stands for *donor* atom. Since each phosphorus atom gives exactly one electron to the silicon crystal, the electron concentration will for all practical purposes be practically equal to the phosphorus concentration. In other words, for a semiconductor doped with phosphorus we have a free-electron concentration:

$$n = N_D \qquad (1.4)$$

This type of doping creates what is called *N*-type material, because the dominant charge carrier is the electron. If an element from column III of the periodic table is used to dope the silicon, an analogous process occurs, but in this case the column III element (boron is the most widely used) has to acquire or accept an electron from the silicon, thus in effect adding more holes to the silicon. The column III element is called an *acceptor* atom. This fact will probably not be immediately evident but we will come back to it in more detail in Chapter 2. However, to complete the picture, we can state that if a material is doped with a concentration of N_A acceptor atoms per cubic centimeter, the hole concentration will be for all practical purposes equal to N_A; thus:

$$p = N_A \qquad (1.5)$$

The material is now referred to as *P*-type material.

1.4 IMPORTANT PROPERTIES AND PARAMETERS OF SEMICONDUCTORS

In order to study electronic devices, there are several properties we need to know (in addition to the band gap energy E_g). For example, the conductivity or resistivity of a solid does not only depend on the concentration (number per unit volume) of electrons in the conduction band. It also depends on the ease with which these electrons can move when an electric field

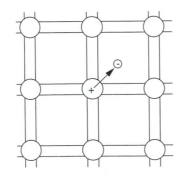

Figure 1.8 Illustration of a donor atom, which gives up its fifth outer-shell electron to become a fixed positively charged ion.

E (note that this E for electric field, units of V/cm, has no connection with E_g for energy band gap, units of electron volt, eV) is applied (by placing a voltage over a certain thickness of semiconductor material). This "ease of movement" is characterized by a parameter called the *mobility*, μ_n, of the electron. The situation shown in Fig. 1.9 is similar to an object moving under friction, or in a viscous medium. In fact, the electron "collides" with lattice atom sites and is scattered, as we will subsequently see.

The electron current density J_n, caused by application of an electric field E, is given by:

$$J_n = q\mu_n n E \tag{1.6}$$

where q is the charge on an electron (1.6×10^{-19} coulombs), n is the electron concentration (electrons per cm^3). This current is referred to as *drift current*.

In many devices, the current that determines the electrical properties is due not to conduction, but to *diffusion* of particles. Just as smoke diffuses through a room from a point of high concentration to a point of low concentration, so electrons can diffuse in a solid from a region of high concentration to a region of low concentration, as indicated in Fig. 1.10. The rate of movement is proportional to the concentration gradient dn/dx, and the corresponding parameter of proportionality is called the *diffusion coefficient*, D_n, which, as we shall later see, is directly related to the mobility. The electron current density due to diffusion is given by:

$$J_n = q D_n \frac{dn}{dx} \tag{1.7}$$

Another important semiconductor property is connected with what happens when electrons are raised to a higher energy level (excited across the band gap) and then allowed to fall back (decay) to the "thermal equilibrium" condition. This does not occur instantaneously when the source of excitation (e.g., light or a voltage applied across a semiconductor diode) is removed, but occurs as an exponential decay, characterized by a time constant called the *lifetime*, τ, of the electrons, as shown in Fig. 1.11. This is the average time an excited electron in the conduction band exists before recombining (across the band gap) with a hole in the valence band and is typically of order microseconds. The carrier lifetime determines the current–voltage relations of many diodes. The excess (above thermal equilibrium) electron concentration $n'(t)$ decay versus time is given by:

$$n'(t) \propto \exp\left(-t/\tau\right) \tag{1.8}$$

The detailed recombination mechanism is discussed in more detail in Section 2.5 and applies also to excess holes.

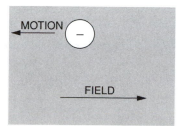

Figure 1.9 Illustration of motion of electron in response to an electric field E. The ease of motion is characterized by the mobility μ_n of the electron.

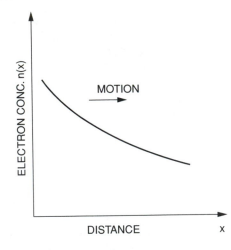

Figure 1.10 Motion due to diffusion of an electron away from a region of high electron concentration.

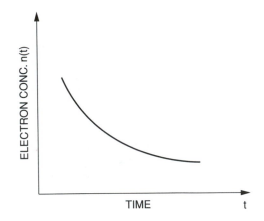

Figure 1.11 Decay of excess electron concentration with time, illustrating the electron lifetime.

1.5 SEMICONDUCTOR DEVICES—A BRIEF OVERVIEW

In this section, we will look very briefly at some of the most common semiconductor devices. In each subsection we start by looking briefly at the device as a circuit element, so that we know how it is used to perform an electronic function; we then discuss briefly how the device is made, or fabricated using silicon technology; finally, for each device, we explain in a purely qualitative manner how the device works. The purpose of the following descriptions is solely in the context of an overview, to give the reader a feel for what each device is and how it is made and used. The circuit behavior aspects by themselves cover a complete undergraduate course, and we will go into no more detail here. Fabrication details are discussed in Chapter 8. The remaining topic, how does the device work?, forms the essential core of the remaining chapters of this text and will be expanded on in considerable detail.

1.5.1 *P–N* junction diode

1.5.1.1 The diode as a circuit element

A diode is an electronic component that ideally passes current in only one direction. The circuit symbol is shown in Fig. 1.12(a) and the corresponding *I–V* characteristic in Fig. 1.12(b).

The *PN* junction diode has an exponential *I–V* characteristic that approximates this unidirectional current flow condition over a wide voltage range:

$$I = I_0 \exp\left(q V_{\mathrm{a}}/kT\right) \tag{1.9}$$

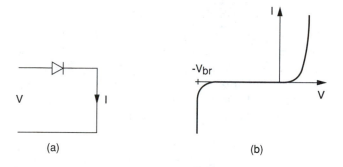

Figure 1.12 (a) Diode symbol with voltage and current. (b) Diode current–voltage characteristic V_{br} is the reverse breakdown voltage.

where I_0 is the reverse leakage current and V_a the applied voltage. For a silicon diode under normal operation, the forward diode drop is of order 0.7 volts. The reverse leakage (saturation) current I_0 is of order 10^{-9} times the forward current, until large reverse voltages are reached. When the bias reaches the reverse breakdown voltage V_{br}, the current increases rapidly in the reverse direction.

Applications

1. *60 Hz (or 50 Hz) rectifier*

This is the most direct way of converting an ac power supply to dc. The simple circuit is shown in Fig. 1.13(a), with the corresponding current waveform in Fig. 1.13(b). The diode requirements are specified by (1) its maximum forward current and voltage drop, and (2) its maximum reverse voltage (the breakdown voltage). These specifications can range from over 100 amps at 0.8 V and a V_{br} of over 1000 volts for high-power diodes to less than 1 amp and V_{br} of order 5 volts for low-power diodes. The diode is often referred to as a *PIN* diode, for reasons to be discussed in Chapter 3.

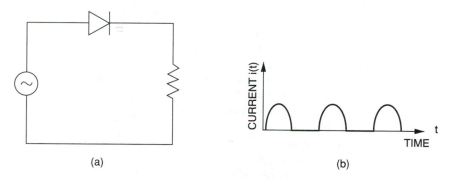

Figure 1.13 (a) Simple diode rectifier circuit. (b) Current versus time waveform of simple diode rectifier.

2. *Limiter, clamp, Zener reference*

Using the reverse breakdown characteristic of a diode enables it to be used as a "limiter" or "clamp," with the breakdown voltage being the main design parameter with applications ranging from 3 to 1000 V. A particular application is when the diode is used as a voltage reference by biasing it at a moderate reverse current to keep it in the breakdown region. This is often referred to as a *Zener diode*, but the true Zener effect only occurs at breakdown voltages less than 6 volts.

3. *Radio frequency (rf) detector*

An amplitude modulated (am) radio signal has to have the radio frequency component removed in order to be amplified and fed to a speaker, where only the low-frequency (lf) component is used. This is done easily by a low-power "detector" diode (actually rectifying the rf signal, but usually referred to as a "detection" process). This diode only has to handle millivolts of signal. In amplitude modulation receivers the detection is often performed by the nonlinear properties of a bipolar transistor. At microwave frequencies, radar signals are often detected by a high-speed diode (one with a very small capacitance and small cross-sectional area)— in fact, this was the first modern semiconductor device, preceding the discovery of the BJT in 1948.

4. *Microwave limiter, switch, attenuator*

This is usually a small (to keep the capacitance low) *PIN* diode and is used under forward bias as a low impedance and at zero or reverse bias as a high impedance in microstrip (e.g., 50 ohm) lines to act as a switch to the microwave signal.

1.5.1.2 How is a diode made?

A semiconductor diode is made by having an *N*-type semiconductor in contact with a *P*-type semiconductor. By far the easiest way to accomplish this is to prepare a good crystalline sample of silicon in the form of a wafer (typically several inches in diameter). The wafer is doped uniformly at a value that ultimately determines the diode breakdown voltage. Let us assume that we have prepared an *N*-type wafer with a uniform donor doping level of about 10^{15} cm^{-3}. If we now put the wafer in a furnace at about 1000°C with a gas containing boron in contact with the wafer surface, the boron atoms will gradually diffuse into the silicon. The boron atomic concentration will be high at the surface, where the solid solubility level is reached. This is of order 10^{21} cm^{-3}. It is useful to note that a typical solid has about 10^{23} atoms per cm^3; so we are talking about a solid solubility limit of order 1 impurity per 100 silicon atoms. The boron atomic concentration falls below the surface, with a roughly Gaussian dependence on depth (proportionally to $\exp[-(x/x_0)^2]$, where x_0 is a characteristic depth critically dependent on the time in the furnace and to the furnace temperature; x_0 is typically of order 1 μm). This is shown in Fig. 1.14(a).

At some depth X_j below the surface, the phosphorus (donor) and boron (acceptor) concentrations are equal. Above this depth the material is *P* type; below this depth the material is *N* type. We thus have a *P–N* junction. The plot shown in Fig 1.14(a) is called the "net impurity concentration," since it represents the magnitude of $N_A(x) - N_D(x)$. Figure 1.14(b) shows a cross section of the diode (i.e., in a plane perpendicular to the silicon surface); from

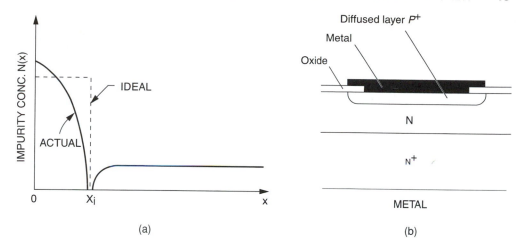

Figure 1.14 (a) Atomic net impurity profile of a diffused *P–N* junction diode. (b) Cross section of silicon *P–N* junction diode.

top down we have the metal contact, the P^+ diffused layer, the background N layer, the highly doped N^+ bottom layer, and the back contact.

In many cases the surface dimensions of the diode are determined by using a *mask* with a photolithographic process to create a *window opening* in a surface *oxide layer*, as shown in Fig. 1.14(b). This allows the boron to diffuse only through the area determined by the mask. The steps involved in this process are described in detail in Chapter 8.

1.5.1.3 How does a diode work?

The essential feature of a *P–N* junction is that the electron concentration is high on the N side and low on the P side. There is a natural tendency for the electrons to flow by diffusion from the N side to the P side. Note that this necessarily means that on the N side there will be a loss of electrons; since each donor atom was electrically neutral to start with, this means that there will be some positive charge on the N side near the junction as seen in Fig. 1.15. The top diagram shows schematically the fixed + (positive) charges due to the ionized donor atoms and the free electron charge, with the opposite situation on the P side, where each acceptor atom has a fixed negative charge, with free positive holes. The lower diagram of Fig. 1.15 is a plot of electron concentration, $n(x)$, and hole concentration, $p(x)$, versus distance.

Under thermal equilibrium conditions (a diode protected from light, with no current flowing), there cannot possibly be any net flow of current. A charge imbalance and a corresponding electric field are established (in Chapter 2 we will use Gauss's law or Poisson's equation to solve for this situation) to oppose the flow of electrons due to diffusion; the net current is zero. This field extending a small distance either side of the actual junction creates a potential barrier, typically of order 0.7 volt in a silicon diode. If a bias voltage is now applied in such a direction as to reduce the barrier, current will flow, mainly by diffusion, with electrons moving from the N side (where they are present in a large concentration) to the P side (where the electron concentration is very small); this is shown in Fig. 1.16(a), where circled charges represent

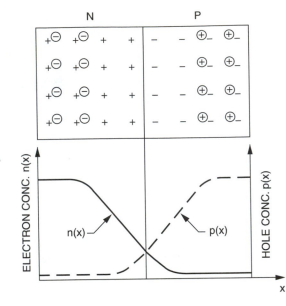

Figure 1.15 *N–P* junction showing fixed impurity atom positive and negative charges and free electrons and holes (circled). The bottom diagram is a plot of $n(x)$, $p(x)$.

"free" or mobile carriers; fixed charges (ionized impurity atoms) are not encircled. This is called *forward* bias. The center region, which has fewer free carriers, is referred to as the *depletion layer* or *space-charge layer*. It will become narrower under forward bias.

If the applied bias voltage is reversed in polarity, so as to increase the potential barrier, corresponding to the situation shown in Fig. 1.16(b), this makes it more difficult for the electrons to flow across and only a very small reverse "leakage" current (typically specified in nA, pA, or fA) flows. In this case the depletion layer increases in width. This leads to the complete current–voltage characteristic of a diode already shown in Fig. 1.12(b). The diode thus acts as a device that conducts current only in one direction and blocks current in the other direction.

The design parameters for a simple rectifier diode are: peak reverse voltage (often called the diode breakdown voltage); maximum forward current (for a given forward voltage of order

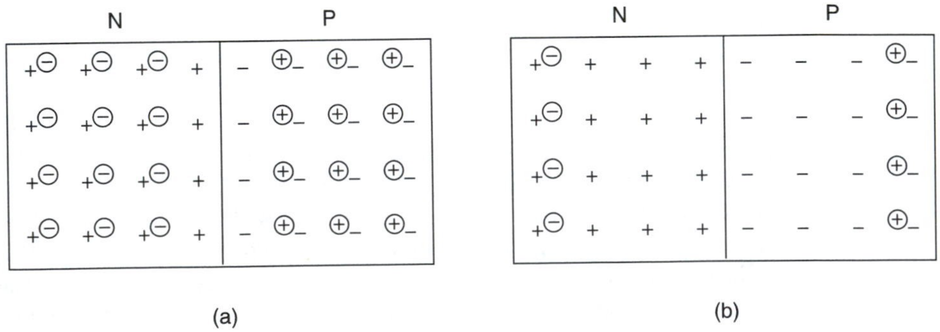

Figure 1.16 (a) Charges in a forward-biased *N–P* junction. (b) Charges in a reverse-biased *N–P* junction.

1 V); speed with which the diode can be turned from the forward to reverse state and vice versa (the diode turn-off and turn-on delay times).

1.5.2 Bipolar transistor (BJT)

1.5.2.1 The BJT as a circuit element

Figure 1.17 shows the circuit symbols for a bipolar transistor, or more explicitly a bipolar junction transistor, or BJT for short. We can have *NPN* or *PNP* devices; the discussion here concerns the *NPN* structure. It is a three-terminal device and is used for three main applications: (1) as an amplifier; (2) as a logic element with on/off states, as in emitter-coupled logic (ECL); (3) as a switch, to control currents as, for example, in electric motor speed control. The three terminals are labeled on the diagram: *base* (normally the input terminal), *emitter* (frequently the common or grounded terminal), the *collector* (normally the output terminal).

The BJT may be connected in three different circuit configurations with the emitter, base, or collector being "common" to input and output. The common-emitter configuration is the most widely used. In this case the BJT has the following properties:

1. A high output impedance (typically tens or hundreds of kohms), which is one factor contributing to a useful level of voltage gain;
2. A high current gain i_c/i_b (of order 100);
3. A high transconductance g_m. This is the incremental output current divided by incremental input voltage. In terms of ac output current i_0 and ac input voltage v_{in}:

$$g_m = i_0/v_{in} \tag{1.10}$$

We follow conventional notation, using lower-case symbols to represent the magnitudes of ac quantities and upper case for dc quantities;

4. A low voltage between collector and base when in the "on" switch or logic state;
5. A very good frequency response (wide bandwidth) and fast switching speed.

In order to use the transistor as a circuit element, it must have a supply voltage V_{CC}, a load or output resistance R_L, and an input bias supply (or an equivalent logic or switched input signal to turn it from the "off" to the "on" state). A typical circuit arrangement is shown in Fig. 1.18; this particular arrangement is called the *common-emitter configuration*; v_0 is the output voltage. The voltage gain $A_v = v_0/v_{in}$ is given by:

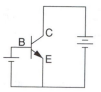

Figure 1.17 Symbol and normal bias for an *NPN* bipolar transistor.

$$A_v = i_0 R_L / v_{in} = g_m R_L \qquad (1.11)$$

Typical values of A_v range from 20 to 100.

The design parameters of major importance for a BJT are: maximum operating voltage, maximum (or optimum) operating current, and cutoff frequency (or switching speed).

1.5.2.2 How is a BJT made?

Fabrication of an *NPN* BJT involves at least one extra diffusion compared to a diode. In cross section (i.e., looking sideways at a transistor cut perpendicular to the wafer surface) it appears as shown in Fig. 1.19. The starting material for a discrete (as opposed to an integrated circuit) BJT is a wafer substrate with high *N*-type doping forming the collector. A moderately doped *N* layer is grown on top. Then a *P–N* junction is created by diffusing boron atoms, as would be the case for a diode. Finally, a smaller *N* emitter region is diffused inside the *P* region.

Contacts are made to this *N* emitter region, to both sides of the *P*-diffused base region, and to the back of the highly doped substrate. The most crucial dimension of the BJT is the vertical base thickness W_b between the two space-charge regions in Fig. 1.19.

1.5.2.3 How does a BJT work?

It is only possible to hint here at how the bipolar transistor works; the details and more convincing arguments will be given in Chapter 4. Consider the bias arrangement of Fig. 1.18 with forward bias between emitter and base, with a reverse bias applied between collector and base terminals (the c–b voltage is significantly greater than V_{BE}, hence ensuring that the c–b junction is reverse biased). The emitter–base junction behaves almost exactly like the

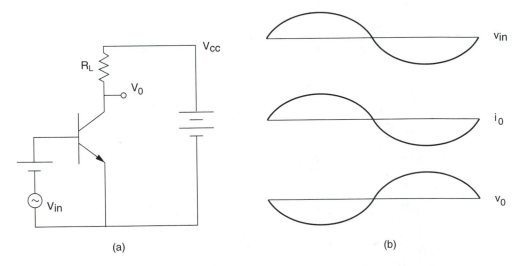

(a)   (b)

Figure 1.18 (a) Typical BJT common-emitter circuit for use as an amplifier or logic inverter; (b) waveforms at input and output.

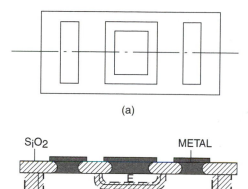

Figure 1.19 (a) Bipolar junction transistor surface (mask) view; (b) cross section perpendicular to wafer surface (the shaded areas are the space-charge layers and the oxide regions). The collector contact is taken from the bottom of the device.

forward-biased diode discussed in Section 1.5.1.3. The forward bias *lowers* the barrier at the emitter–base (e–b) junction and allows electrons to flow toward the base. However, the base is made very thin (typically much less than 1 μm in a modern BJT), and the reverse-biased collector junction is very close to the emitter. So almost all the electrons flow across the base and through the base–collector junction and out the collector terminal. It is most important to realize that although the collector–base junction is reverse biased; this reverse bias actually *helps* the electrons to flow. This can be understood most readily by thinking about the battery and the direction of current flow in the external circuit of Fig. 1.18. The V_{CC} battery voltage will make conventional current flow from $+$ to $-$ in the external circuit; that is, it will make electrons flow *out* of the collector into the battery. This voltage is quite clearly and unambiguously helping the electrons to leave the base and go through the collector. There is no contradiction here. Typical dc characteristics of an *NPN* BJT are shown in Fig. 1.20.

The BJT is characterized by its common-emitter current gain $\beta = I_C/I_B$, which remains roughly constant for a wide range of bias conditions and is typically of order 100. The collector current is an exponential function of base–emitter voltage, as in a diode:

$$I_C = I_{CS} \exp\left(q V_{BE}/kT\right) \tag{1.12}$$

The transconductance defined by Eq. (1.10) is [see Fig. 1.20(b)]:

$$g_m = i_0/v_{in} = \frac{dI_C}{dV_{BE}} = \frac{I_C}{(kT/q)} \tag{1.13}$$

and is thus proportional to the dc collector current.

1.5.3 Junction field-effect transistor (JFET)

1.5.3.1 The JFET as a circuit element

The circuit symbol for a junction field-effect transistor or JFET is shown in Fig. 1.21. We can have *N*-channel or *P*-channel devices; the discussion here is for an *N*-channel device.

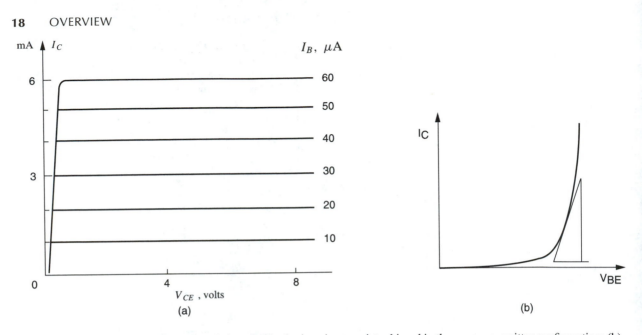

Figure 1.20 (a) Ideal $I_C - V_{CE}$ characteristics of a bipolar junction transistor biased in the common-emitter configuration; (b) $I_C - V_{BE}$ characteristics showing the slope for g_m calculations.

Like the BJT the JFET is a three-terminal device used for the same purposes as the BJT. In normal operation, the input is referred to as the *gate*; the common grounded terminal is the *source*; the output is normally the *drain* terminal.

The main features of the JFET are as follows:

1. A very high input resistance;
2. A moderately high output impedance;
3. A rather low transconductance;
4. A fair to good frequency response and switching speed.

Regarding item (4), it is worth noting that the GaAs MESFET (metal semiconductor FET) is a very high-speed device that works on almost identical principles to the JFET.

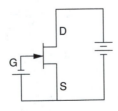

Figure 1.21 Junction field-effect transistor (JFET) symbol and normal bias for *N*-channel device.

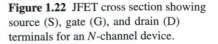

Figure 1.22 JFET cross section showing source (S), gate (G), and drain (D) terminals for an *N*-channel device.

Design parameters for the JFET are: maximum operating voltage, maximum operating current, frequency response/switching speed.

1.5.3.2 How is a JFET made?

The JFET appears somewhat simpler than the BJT in terms of its fabrication. Let us consider an ideal structure as shown in Fig. 1.22. The wafer in this example is low-doped *P* type. On top of this is grown an *N* layer of moderate doping level. A *P–N* junction is now formed by diffusing boron from the top surface. Contacts are made to the diffused boron region (the gate), and to the *N* region on either side of the gate (the source and drain).

1.5.3.3 How does a JFET work?

Operation of the JFET is simpler to understand than that of the BJT, provided we can accept one basic property of *P–N* junctions previously only briefly mentioned. We will explain this property here, without any attempt at a formal proof. Because of the potential barrier set up when a *P–N* junction is made, and to the flow of electrons from the *N* side to the *P* side, it is (perhaps) clear based on the discussion in Section 1.4.1.3 that a *space-charge layer* is created on either side of the junction. This is shown in Fig. 1.23. The electrons that have left the *N* side leave behind a net positively charged region, depleted of the original electrons supplied by each donor atom. A similar space-charge or depletion region exists on the *P* side. It is not too difficult to imagine that the extent (i.e., the thickness) of this space-charge region can be varied by altering the applied bias voltage; increasing the reverse bias will increase the space-charge layer thickness; increasing the forward bias will decrease its thickness, as described in Section 1.5.1.3. By examining the space-charge region underneath the *P*-diffused layer of the JFET shown in Fig. 1.24, we see that its thickness can be altered by changing the bias voltage applied between the source and gate contacts.

Now comes the crucial part of the explanation. As the reverse voltage increases, thus increasing the space-charge region thickness, so the thickness of the conducting *N-channel* region is decreased (current cannot flow in the space-charge region since there are no free carriers there). In other words, the resistance between the source and drain terminals can be altered,

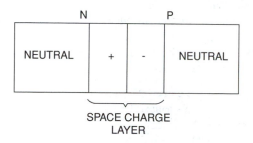

Figure 1.23 Creation of a space-charge layer on either side of an *N–P* junction.

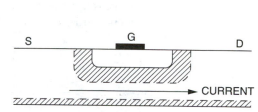

Figure 1.24 Cross section of a JFET showing space-charge regions, with current indicated by the arrow flowing horizontally between the two space-charge layers.

because we now have a means of altering the thickness of the conducting region, that is, of altering the cross-sectional area A_s of the resistance $R = \rho L/A_s$, where ρ is the channel resistivity and L the length in the direction source–drain. For a given drain voltage, this means that the drain current can be altered by changing the gate to source bias voltage. We thus have a device with transconductance. For normal amplifier operation, the drain–source voltage is kept high enough to ensure a high output resistance, the dc characteristics being as shown in Fig. 1.25.

The JFET is characterized by the maximum drain current I_{DSS}, the pinch-off voltage, V_P, the voltage at which the $V_{GS} = 0$ curve becomes flat or saturates, which is also the voltage V_{GS} required to reduce the current I_{DS} to zero. The maximum drain current is given by:

$$I_{DSS} = V_P/3R_{CO} \qquad (1.14)$$

where R_{CO} is the resistance of the horizontal channel at zero bias given by $\rho L/aZ$, where a is the vertical thickness of the conducting channel and Z is the width perpendicular to the page in Fig. 1.24.

The transconductance in the saturated region of the characteristics is given by:

$$g_m = (1/R_{CO})[(V_{SG}/V_P)^{1/2} - 1] \qquad (1.15)$$

It is useful to note that the maximum drain current and maximum transconductance are hence related by:

$$g_m = 3I_{DSS}/V_P \qquad (1.16)$$

By comparing with Eq. (1.13) for the BJT, it is seen that for a given dc current, the JFET has a transconductance that is smaller in the ratio $3(kT/q)/V_P$. Since $kT/q = 0.0259$ V at room temperature, and V_P is typically of order 5 V, the JFET has a maximum transconductance, which is typically of order 0.015 times that of a BJT for the same current.

1.5.4 MOS field-effect transistor (MOSFET)

1.5.4.1 The MOSFET as a circuit element

The symbol for a metal oxide semiconductor field-effect transistor or MOSFET is shown in Fig. 1.26(a), with a simple amplifier circuit in Fig. 1.26(b). We will discuss the N-channel device here. Not surprisingly, the device has very similar characteristics to the JFET. The main difference is that the dc or low-frequency input resistance is now virtually infinite because the input gate is in series with a thin oxide layer, providing near-perfect insulation properties. The overall properties of the MOSFET as a circuit element are thus:

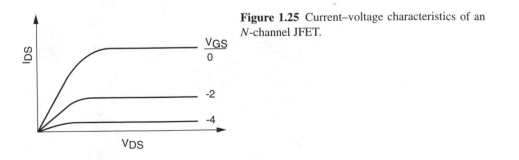

Figure 1.25 Current–voltage characteristics of an N-channel JFET.

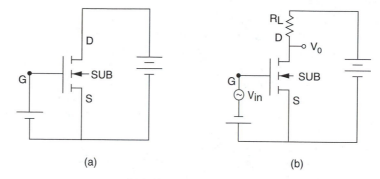

(a)   (b)

Figure 1.26 (a) Metal oxide semiconductor field-effect transistor (MOSFET) symbol and bias arrangement for an N-channel device; (b) common-source amplifier circuit with load resistance R_L and source voltage v_{in}.

1. Infinite dc input resistance;
2. Moderately high output resistance;
3. Moderate transconductance;
4. Moderate frequency response or switching speed.

N and P-channel MOSFETs may be combined to form a CMOS inverter. This finds widespread use today in VLSI circuits because of its ease of fabrication and low standby power consumption.

1.5.4.2 How is a MOSFET made?

The fabrication of an N-channel MOSFET appears very simple, as shown in Fig. 1.27. Starting with a lightly doped P wafer substrate, a thin oxide region of order 0.1 μm (1000 Å, 100 nm) thick is created in what will be the gate area. N diffusions are created for the source and drain areas. Three contacts complete the (misleadingly simple) picture of how a MOSFET is made. Nevertheless, the MOSFET is intrinsically easier to fabricate than a BJT or even a JFET, which was one of the reasons for its early success as an active semiconductor device. The main reason for its use today is not for ease of fabrication (which for CMOS is actually very complex), but because of its power advantage and to a certain extent its IC design simplicity in CMOS logic applications.

1.5.4.3 How does a MOSFET work?

This is perhaps the most difficult device to explain (although some people would argue that the bipolar transistor is more complicated). We shall attempt a purely qualitative explanation

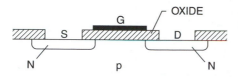

Figure 1.27 Cross section of an N-channel MOSFET. Oxide regions are shown shaded.

at this stage. In a conventional capacitor, if a positive charge is placed on one metal plate, a negative charge is induced on the other plate (overall charge balance must be retained), as shown in Fig. 1.28. Now let us look at the cross section of the MOSFET oxide gate region as shown in Fig. 1.27.

The N-type source and drain diffusions are interconnected by the P substrate. But this means that we have an NP junction in series with a PN junction; so no current will flow when a positive voltage is connected between drain and source (or rather, the current will be that of a reverse-biased diode, of order nA). If a positive voltage is now placed on the metal gate, and if this voltage is sufficiently high (typically a couple of volts), a negative charge is induced on the semiconductor surface just under the gate oxide. This negative charge is made up of electrons. We now have the N source and drain regions with an electron path between them; so current can now flow. In other words, we can control the drain–source current by varying the gate voltage. The device exhibits transconductance, with near-infinite input resistance, and moderately high output resistance.

The dc characteristics of a typical MOSFET are shown in Fig. 1.29. These are shown for an "enhancement-mode" device, the most common type of MOSFET. Note that the output current only starts to flow when the input gate voltage is above some "threshold value" V_{th}, and the current then increases as the input voltage is increased. The drain current in the flat or saturated part of the $I_{DS} - V_{DS}$ characteristic (i.e., for $V_{DS} > V_{GS} - V_{th}$) is given by:

$$I_{DS} = (Z/L)\mu C'_{ox}(V_{GS} - V_{th})^2/2 \tag{1.17}$$

where Z is the channel width perpendicular to the page in Fig. 1.27, L is the channel length between source and drain, μ is the carrier mobility, C'_{ox} is the gate oxide capacitance per unit area ϵ_{ox}/t_{ox}; the value of the permittivity of SiO_2 is 0.3×10^{-12} F/cm. The oxide thickness t_{ox} is a critical parameter in MOSFET fabrication and lies typically in the range 100 to 1000 Å (10 to 100 nm). This is shown in Fig 1.29(b).

The small-signal transconductance in the flat or saturated region of the $I_{DS}-V_{DS}$ characteristics is given by the slope of the I_{DS} versus V_{GS} curve, as indicated in Fig. 1.29(b):

$$g_m = \frac{dI_{DS}}{dV_{GS}} = (Z/L)\mu C'_{ox}(V_{GS} - V_{th}) \tag{1.18}$$

Combining Eqs. (1.17) and (1.18), we see that the output dc current and transconductance are related by:

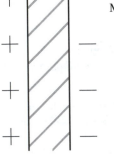

Figure 1.28 Charge on a conventional capacitor or on the gate region of a MOSFET.

$$g_{m} = 2I_{DS}/(V_{GS} - V_{th}) \tag{1.19}$$

Comparing with the BJT, we see that the ratio of the transconductance for the MOSFET compared to that of the BJT at the same dc current is $2(kT/q)/(V_{GS} - V_{th})$. For a typical bias condition of $V_{GS} = 4$ V for $V_{th} = 2$ V, the MOSFET transconductance is thus smaller by factor of order 0.025.

The low transconductance of the JFET and MOSFET has implications when circuit load capacitances are high, since the associated charging delay can be significant.

1.5.4.4 The CMOS inverter

Since the main application of MOSFET devices today is in their use for CMOS integrated circuits, we show in Fig. 1.30 how an *N*-channel (driver) and a *P*-channel (load) device are connected to form a CMOS logic inverter. This inverter possesses the very attractive property of extremely low standby current (of order nA). This means that at low clock rates, or for static memory cells, the power consumption can be kept very low.

1.5.5 Other semiconductor devices

There are a very large number of other semiconductor devices. The devices to be covered briefly in Chapter 7 are as follows:

OPTOELECTRONIC DEVICES. The photodiode is a *PN* junction, normally reverse biased and used as a detector of optical signals, for example, from a fiber optic link. The band gap determines the optical cutoff wavelength, and the other main design parameters are the sensitivity (limited by internal noise sources) and bandwidth. The solar cell is a special case, where the *PN* diode parameters are optimized to give maximum electrical output power for incident sunlight. Light-emitting diodes and semiconductor lasers form a separate group of

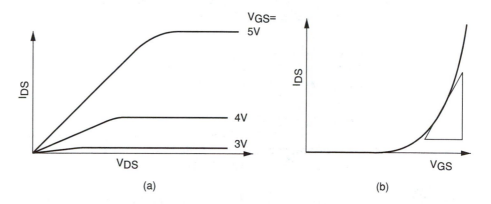

Figure 1.29 Typical MOSFET current–voltage characteristics: (a) I_{DS} vs. V_{DS} for various V_{GS} values; (b) I_{DS} vs. V_{GS} in the saturation region ($V_{DS} > V_{GS} - V_{th}$), showing the slope for g_{m} calculations. I_{DS}, V_{DS}, V_{GS} are positive for *N*-channel enhancement MOSFETs and negative for *P*-channel devices.

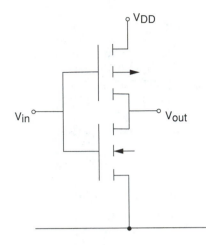

Figure 1.30 CMOS inverter using an *N*-channel and a *P*-channel MOSFET.

optoelectronic devices and will be introduced. Finally, an overview of the charge-coupled device (CCD) for imaging array applications will be given.

THE THYRISTOR OR SILICON-CONTROLLED RECTIFIER. This is a very high-voltage switching device, consisting essentially of two overlapping BJT structures. The design parameters are: maximum rated voltage (typically in excess of 1000 V), maximum current, switching speed.

HIGH-POWER MOS DEVICES. These include the DMOS and VMOS structures that use special techniques to parallel a large number of individual MOSFETs to increase the current handling ability.

THE INSULATED GATE BIPOLAR JUNCTION TRANSISTOR (IGBT). This is a modern high-voltage switch with the advantages of MOSFET input resistance and BJT current driving ability. Design parameters are the same as for the thyristor.

MICROWAVE DEVICES. These cover a wide range of *P–N* junction devices whose applications range from microwave switches, attenuators, frequency multipliers, varactor tuning diodes, to IMPATT and transferred electron structures.

THE HETEROJUNCTION BIPOLAR TRANSISTOR (HBT). This BJT structure makes use of different band gaps in emitter and base regions, or across the base, to obtain better high-frequency performance. The main materials used at present are SiGe (for the base, with a silicon emitter and collector) and GaAlAs (emitter) /GaAs (base).

MOS MEMORY ELEMENTS. These include the FAMOS floating-gate structure, EEPROMs, and similar devices that retain charge in an insulated gate for many years, thereby enabling the device to be set to normally "on" or normally "off" and hence to be used in memory arrays.

1.6 CONCLUSIONS

In this chapter we outlined the basic properties of semiconductor material and introduced the concepts of energy bands, band gap, and electrons and holes as particles that carry current. We explained the manner in which the carrier concentration can be controlled by doping the semiconductor with donor or acceptor atoms; the two basic current mechanisms—drift and diffusion—were explained, and the concept of recombination was mentioned. The last part of the chapter gave an overview of diodes, BJTs, JFETs, and MOSFETs; this survey included the elementary circuit properties of each device, its construction, and how it works.

We conclude that the *PN* diode depends on injection of holes and electrons, as does the bipolar transistor. The JFET and MOSFET, on the other hand, depend on only one type of carrier (electrons for the *N*-channel JFET and *N*-channel MOSFET) for their operation. They are, therefore, sometimes referred to as unipolar devices or majority carrier devices.

Subsequent chapters will expand on these items, going into considerably more detail concerning the internal operation of each device.

PROBLEMS

1. For a hydrogen atom, what is the energy (in eV and joules) required to raise an electron from the second to the third energy level? What is the wavelength of light (in microns) that would be emitted if this electron falls back to (a) the second energy level; (b) the ground state?

2. Calculate the maximum wavelength (in microns) at which light can be detected (using a *p–n* junction diode or change in conductivity) in the materials listed in Table 1.1. Hence state which materials could be used for detection in the visible spectrum. In fiber optic systems, minimum attenuation occurs at approximately 1.3 μm; which material could be used for a fiber optic detector?

3. If the spacing of atoms in a silicon crystal lattice as shown in Fig. 1.7 is approximately 4 Å, calculate the number of atoms in one cubic centimeter of silicon. If one phosphorus atom is added for every million silicon atoms, what is the phosphorus atomic concentration per cubic centimeter? If the mobility of electrons in silicon is 1000 cm^2/V s, calculate the resistivity ρ of this doped material? (Hint: $1/\rho = \sigma = q\mu N_D$.)

4. Using the result of Problem 3, estimate the number of atoms in the world (diameter approximately 12,000 km), assuming silicon as a typical element.

5. Assuming that the concentration of free carriers n_i in a pure semiconductor is proportional to $\exp(-E_g/2kT)$, calculate at 300 K: (a) the factor by which n_i is reduced in carbon ($E_g = 5.5$ eV) compared to silicon; (b) the factor by which n_i is increased in germanium ($E_g = 0.7$ eV) compared to silicon.

6. If a pure silicon sample whose free carrier concentration n_i is 10^{10} cm^{-3} is doped uniformly with phosphorus atoms to a concentration of 10^{16} cm^{-3} and simultaneously with an equal uniform concentration of boron atoms, explain why the free carrier concentration is still 10^{10} cm^{-3}.

7. Calculate the magnitude of electron drift current density in silicon doped 10^{15} cm^{-3} for an electric field of (a) 1 V/cm; (b) 1000 V/cm (assume $\mu_n = 1000$ cm^2/V s).

8. Calculate the magnitude of electron diffusion current density if an electron concentration of 10^8 cm^{-3} falls to a value of 0 in a distance of 1 μm. Assume $D_n = 30$ cm^2/s.

9. If the carrier lifetime of electrons in silicon is $\tau = 10\,\mu s$, calculate the time for the electron concentration to fall to (a) 50%; (b) 1% of its initial value after a light source is removed.

10. In a P–N diode, if the surface concentration of boron is 10^{21} cm^{-3} and the boron distribution is gaussian with a characteristic length $x_0 = 1\,\mu m$, calculate the junction depth X_j for (a) a background doping of 10^{17} cm^{-3}; (b) a background doping of 10^{15} cm^{-3}.

11. A diode has a reverse leakage current $I_0 = 1$ nA at 300 K. Assuming an ideal I–V law,

    (a) Calculate the forward current for an applied forward voltage V_a of: (i) 0.6 V; (ii) 0.8 V.

    (b) If the circuit is such that the current is 1 mA, calculate the value of voltage V_a.

    (c) Calculate the reverse current for $V_a = -0.1$ and -1 V.

12. A bipolar transistor is used in the circuit of Fig. 1.18 with $V_{CC} = 10$ V, $I_C = 5$ mA, $R_L = 1$ kohm. Calculate (a) the dc voltage on the collector; (b) the maximum possible ac output peak-to-peak voltage swing; (c) the transconductance; (d) voltage gain; (e) the required input peak-to-peak voltage for the above output voltage. If $I_{CS} = 10^{-15}$ A, calculate the required value of dc base–emitter bias V_{BE}.

13. For the BJT example of Problem 12, calculate peak maximum and minimum values of the output voltage waveform if the input ac voltage v_{in} is a sine wave with peak values of (a) 1 mV; (b) 20 mV; (c) 100 mV. Sketch the output waveform for the three cases.

14. An N-channel JFET has a zero-bias channel resistance $R_{CO} = 5$ kohm and a pinch-off voltage $V_P = 5$ V. Calculate the maximum transconductance for a gate–source bias $V_{GS} = -1$ V. If this JFET is used in the circuit of Fig. 1.21 but with a load resistance $R_L = 50$ kohm inserted between the V_{DS} battery and drain contact (similar to Fig 1.18 for the BJT), calculate the voltage gain.

15. An N-channel metal gate MOSFET has gate dimensions $1 \times 1\,\mu m^2$. The gate oxide thickness is $t_{ox} = 0.1\,\mu m$. Calculate the gate capacitance (the permittivity of SiO$_2$ is 0.3×10^{-12} F/cm).

16. For the MOSFET of Problem 15, calculate the transconductance at $V_{GS} = 5$ V if the threshold voltage $V_{th} = 2$ V. Assume a channel mobility of 500 cm^2/V s. How does this transconductance compare to the BJT transconductance of Problem 12? Assume that L is kept at 1 μm and calculate the value of Z necessary to have the same transconductance as the BJT in Problem 12.

17. This problem concerns the comparison of a BJT and a MOSFET inverter with a supply voltage V_{supply} (V_{DD} or V_{CC}) of 5 V and a load capacitance $C_L = 1$ pF connected from the output node to ground. Consider the BJT of Problem 12 with $R_L = 1$ kohm and the MOSFET of Problems 15 and 16 with $L/Z = 1$. A voltage step is applied at the input just sufficient to reduce the output voltage to approximately zero. Calculate: (a) the required load resistance for the MOSFET inverter for $V_{GS\,max} = V_{supply}$; (b) the time t_c taken in both BJT and MOSFET cases to discharge this capacitance from V_{supply} to 0 V (assume constant charging current and neglect internal delay times of the transistors). *Hint:* use $Q = CV = i_0 t_c$.

References

1. W. C. Dunlap, *An Introduction to Semiconductors*. New York: Wiley, 1975.
2. A. Bar-Lev, *Semiconductors and Electronic Devices*. 3rd Ed., New York: Prentice-Hall, 1993.
3. A. S. Sedra, K. C. Smith, *Microelectronic Circuits*. Orlando, Florida: Holt, Rinehart, Winston, 1991.
4. D. H. Navon, *Semiconductor Microdevices and Materials*. New York: Holt, Rinehart, Winston, 1986.
5. D. A. Neamen, *Semiconductor Physics and Devices*. Chicago: Irwin, 1997.

Chapter 2

Material Properties
and Basic *P–N* Junction Relations

2.1 INTRODUCTION

In this chapter we examine some basic properties of semiconductor materials. First, for *uniform* doping and *constant* electric field, we investigate electron and hole concentrations, mobility, diffusion coefficient, and drift velocity; second, for *nonuniform* doping (the extreme case of which is the abrupt *P–N* junction) we investigate the existence of space-charge regions and some of the properties of these regions obtained by solving Poisson's equation. We start by recalling the basic band structure of a semiconductor. Some background for this is given in Appendix F, but as stressed in Chapter 1, since the band gap and related properties can all be *measured*, the quantum mechanical theory is not essential for accepting the more important basic properties of semiconductors. The short overview included in Appendix F is made available to convince those readers who need convincing that the band model and related properties are acceptable starting points.

2.2 THE SEMICONDUCTOR IN THERMAL EQUILIBRIUM

2.2.1 Intrinsic (pure) semiconductor material

Let us recall the band diagram introduced in Chapter 1 and reproduced here as Fig. 2.1. E_g is the difference in energy (expressed conveniently in electron volts, eV, the energy required to raise an electron through a potential of 1 volt) between the valence band, which is almost completely filled with electrons, and the conduction band, which is almost empty, at room temperature in pure or intrinsic semiconductor. By the valence band being "filled" we mean that all the available quantum states are occupied. This means that there are no "positions" to which the electrons can move when, for example, an electric field is applied. Consequently no current can flow. In an *almost* filled band some quantum states are not occupied; so some current can flow, as we observed in Chapter 1. It is more convenient mathematically in this

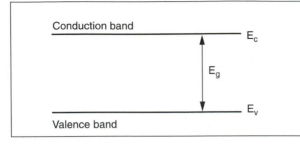

Figure 2.1 Band-gap diagram for a semiconductor: E_c is the conduction-band edge, E_v is the valence-band edge, E_g is the band gap.

case to use the concept of "holes" as unit positively charged particles with an effective mass similar (but not identical) to the mass of an electron.

Concept of the hole as a "particle"

Consider Fig. 2.2, where we show a "box" with five squares, with an electron in each square but the first. Consider an electric field applied so that each of the four electrons moves one space to the left. Consider now a second set of boxes, with one positively charged particle as shown. Applying the same electric field, this one positive charge can be moved four spaces to the right. Then we have the same total movement of charge as in the first situation. If the movement occurs when opposing some force (e.g., similar to friction), then the total amount of work done will be the same in both cases.

In the valence band of the semiconductor, which is nearly full (of electrons), it is clearly possible (and preferable) to consider the movement of a small number of positive charges instead of a very large number of electrons.

In both cases (electrons in the conduction band, holes in the valence band), the effective mass (referred to as m_e^*, m_h^*) must be used. The effective mass represents the electron (or hole) as a classical particle obeying Newton's laws of motion. It differs slightly from the rest mass of an electron in free space because of the fact that the electron in the semiconductor interacts with the forces due to the crystal lattice potential (see Appendix F).

In an intrinsic semiconductor (one to which no dopant atoms have been added), the number of electrons per unit volume in the conduction band (the free-electron concentration) is equal to the number of holes in the valence band. This is expressed by:

$$n_0 = p_0 = n_i \tag{2.1}$$

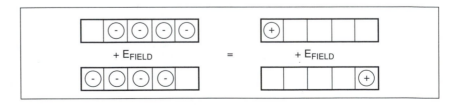

Figure 2.2 Simplified analogy of the hole as a particle.

where the subscript 0 is used to denote thermal equilibrium conditions and where n_i is called the intrinsic carrier concentration. The band gap E_g determines the free-carrier concentration in an undoped semiconductor. The larger the band gap, the smaller are the values of n_0, p_0. Since in thermal equilibrium, the only energy available is the thermal energy of the particles (of order kT, where k is Boltzmann's constant $= 1.38 \times 10^{-23}$ joules per K), it follows that if E_g is much higher than kT, very few electrons will be excited from the valence band to the conduction band. In fact, a rigorous solution (of which a summary is given in Appendix F) gives the following result for an intrinsic semiconductor:

$$n = p = n_i = C \exp(-E_g/2kT) \tag{2.2}$$

where C is a constant. It is shown in Appendix F that the constant is in fact equal to $\sqrt{N_c N_v}$, where N_c, N_v are called the effective density of states; at room temperature in silicon $N_c = 2.8 \times 10^{19} \text{cm}^{-3}$, $N_v = 1.04 \times 10^{19} \text{cm}^{-3}$. In silicon, $E_g = 1.12$ eV, and since $kT = 0.0259$ eV at room temperature (300 K), the free-electron and -hole concentrations are of order 10^{10} cm^{-3} (published values vary from 1.2×10^{10} to $1.6 \times 10^{10} \text{cm}^{-3}$ at 300 K), a very low concentration, making intrinsic silicon almost like an insulator at room temperature. Note that it is not necessary to derive the above result mathematically; it can be observed experimentally, by measuring the conductivity of the semiconductor over a range of temperatures and plotting conductivity ($\sigma \propto n$) or electron concentration, n, on a logarithmic scale versus reciprocal temperature. Provided the semiconductor has a low enough doping level, or that the measurements are carried out at sufficiently high temperatures, the high-temperature rising part of the curve of the appearance of Fig. 2.3 is obtained (typically for $T > 500$ K).

The slope of the curve at high temperatures gives the band gap E_g (from E_g/k), and the value at any temperature may be used to extract the pre-exponential constant (providing the appropriate mobility value has been determined separately). If the semiconductor is doped, the conductivity tends to a constant value at intermediate temperatures and falls off at very low temperatures when kT falls far below the ionization energy of the donor atoms (see the following).

It may be noted that using thermal energy to excite electrons from the valence band across the band gap (thereby creating more electrons in the conduction band and simultaneously more holes in the valence band) is the same as breaking covalent bonds in the bond model discussed in Chapter 1. If there is no other source of free carriers, the electron and hole concentrations must always be equal.

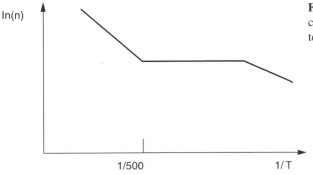

Figure 2.3 $\ln(n)$ versus $1/T$ showing constant gradient $-(E_g/2k)$ for high temperatures.

Example: A semiconductor sample has a resistivity ρ at 150, 200, 250°C equal to 350, 51, 11 ohm cm, respectively. Since $\rho = 1/q(\mu_n + \mu_p)n_i$, we can write (assuming constant mobility):

$$\rho(200)/\rho(250) = n_i(250)/n_i(200) = \exp[-E_g/2k(273 + 250)]/\exp[-E_g/2k(273 + 200)]$$

$$51/11 = \exp\{-(E_g/2)[1/(523k) - 1/(473k)]\}$$

Note that it is often simpler to recall that $kT = 0.0259$ eV at 300 K than to look up the value for k and get the right units (eV or joules); i.e., in the above equation, $523k = 0.0259 \times 523/273 = 0.0496$ eV and $473k = 0.0449$ eV. We thus have:

$$\ln(51/11) = -E_g(10.08 - 11.14)$$

$$E_g = 1.534/1.06 = 1.44 \text{ eV}$$

The third temperature and resistivity are necessary only to ensure that the resistivity is not determined by a background doping level. In fact, with more data, one can also extract the background doping level and even estimate the mobility. See Problem 2.15.

2.2.2 The doped semiconductor

If donor atoms from column V of the periodic table (such as phosphorus) are added to the semiconductor, with each donor atom contributing one extra electron, the electron concentration is increased above the n_i value of the intrinsic crystal. There are now three sources of electric charge present: negatively charged electrons, positively charged holes, and positively charged fixed donor atoms (ions). The fixed charges exist because each donor atom has given up its electron. If the donor concentration N_D is high enough, we can neglect the charge due to holes and write:

$$n_{n0} \approx N_D \qquad (2.3)$$

where N_D is the donor concentration per cubic centimeter and where we have introduced the subscript n to denote N-type material (material doped with donor atoms) and the subscript 0 to denote thermal equilibrium conditions.

Mass action law

A very important relationship exists between the electron and hole concentrations in the presence of impurity (dopant) atoms. Thermal energy excites electrons from the valence band up to the conduction band, as illustrated in Fig. 2.4. There is a natural tendency for the electrons in the conduction band to recombine with the holes in the valence band. The recombination probability is higher if there is a large number of vacancies (holes) to be occupied by the electrons. It is not difficult to accept that the recombination rate is proportional to the np product (i.e., the product of the electron concentration times the hole concentration). Since in thermal equilibrium the thermal generation and the recombination processes must balance, it follows that the thermal generation rate (which is independent of the free carrier concentration)

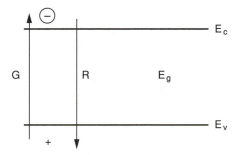

Figure 2.4 Illustration of generation and recombination across the band gap.

$G(T)$ = recombination rate, $R(T) = Cnp$; in other words, the np product is constant at a given temperature and must be equal to n_i^2 for undoped material, hence giving:

$$np = n_i^2 \tag{2.4}$$

From the previous result for the value of n and p in intrinsic material, it follows that we can express the mass action law as:

$$np = N_c N_v \exp(-E_g/kT) \tag{2.5}$$

Note that the temperature dependence of N_c and N_v in Appendix F is negligible compared to the exponential dependence on $1/T$. A typical doping level in a silicon device is $N_D = 10^{16} \text{cm}^{-3}$. Since there are approximately 10^{23} atoms per cubic centimeter in a crystal, this doping level corresponds to only one impurity atom per 10^7 silicon crystal atoms. Since at room temperature $n_i = 10^{10} \text{cm}^{-3}$, it is clear that for all practical purposes the electron concentration, n_{n0} in uniformly doped N-type material is equal to the donor atom concentration, N_D. The exact expression from charge neutrality considerations is:

$$n_{n0} = N_D + p_{n0} \tag{2.6}$$

The hole concentration is then for all practical purposes given by

$$p_{n0} = n_i^2/N_D \tag{2.7}$$

$$\approx 10^4 \, \text{cm}^{-3}$$

in this example (taking $n_i = 10^{10} \, \text{cm}^{-3}$).

If column III atoms (with three electrons in the outer shell such as boron) are added to the intrinsic semiconductor crystal, the reverse process occurs. Each acceptor atom takes an electron from the crystal to form the covalent bond with the silicon lattice, thereby introducing a free hole. The acceptor atom is now a negatively charged ion at a fixed location in the crystal. For uniformly doped *P*-type material, where we use the subscript *p*, charge neutrality enables us to write:

$$p_{p0} = N_A + n_{p0} \tag{2.8}$$

For all practical doping levels we can say that $p_{p0} = N_A$ and that the electron concentration in *P*-type material is equal to n_i^2/N_A in thermal equilibrium.

Ionization energy of a donor atom (to free its fifth outer orbit electron)

A phosphorus atom imbedded in Si may be thought of as similar to the Bohr model of the hydrogen atom with its one orbiting electron, except that the relative dielectric constant ϵ_r of the material is now that for silicon instead of free space and the effective mass of the electron m_e^* should be used instead of m_0. The Bohr energy formula for $n = 1$, from Chapter 1 is:

$$E_h = \frac{m_0 q^4}{8\epsilon_0^2 h^2} = 13.6\,\text{eV} \tag{2.9}$$

Using $\epsilon_r = 12$ for silicon and $m_e^*/m_0 = 0.5$ gives the ionization energy E_{di} of the donor electron:

$$E_{di} = E_h \left(\frac{m_e^*}{m_0}\right)\left(\frac{1}{\epsilon_r}\right)^2$$
$$= 13.6 \times 0.5 \times \left(\frac{1}{12}\right)^2 \tag{2.10}$$
$$\approx 0.05\,\text{eV}$$

Since each particle (electron) has a thermal energy of order $kT = 0.025$ eV at room temperature, it is easy to accept that a large fraction of the donor atoms will be "ionized," that is, will have given up their extra electron. The actual fraction is given without proof in Appendix F (F.21a) and is close to 100% at room temperature and moderate doping levels. We therefore assume normally that: $n \approx N_D$. Likewise for *P*-type material we can define an ionization energy for holes E_{ai} and by analogy conclude that for moderate *P*-type doping levels $p = N_A$ Eq. [(F.21b) gives the exact result].

Note that for $N_D = 10^{16}$ cm^{-3}, the free electron concentration in Si at room temperature has increased by one million compared to intrinsic silicon, but the crystal still has only one impurity per ten million crystal atoms. The conductivity σ increases proportionately to the electron concentration n, that is, by a factor of order 10^6 using $n_i = 10^{10}$ cm^{-3}.

Heavy doping effects

When the doping level is very high, some important effects come into play. The theory for these effects is very complex; however it is a simple matter to estimate the doping level at which one can expect significant deviations from otherwise "normal" laws. Let us again consider the

Bohr model for the donor electron orbiting the donor atom. The radius of this orbit can again be be calculated using the hydrogen atom model, with the appropriate values for dielectric constant and effective mass. We find:

$$r_d = \epsilon_0 h^2 / (q^2 \pi m_0) = 0.53 \times 10^{-8} (\epsilon_r / \epsilon_0)(m_0 / m_0^*) \text{ cm}$$
$$= 13 \times 10^{-8} \text{ cm}$$

$$(2.11)$$

Consider now the situation shown in Fig. 2.5, where as many donor atoms are "packed in" so that the orbitals of the fifth electrons just start to overlap. The critical donor impurity concentration N_{crit} at which this occurs is given by calculating the volume of the cube V_d corresponding to the orbit of one electron;

$$N_{crit} = 1/V_d = 1/(2r_d)^3$$
$$= 1.7 \times 10^{19} \text{ cm}^{-3}$$

$$(2.12)$$

At a value somewhat below this donor concentration, the orbiting electrons start to interact with one another (the Pauli exclusion principle states that not more than two electrons of opposite spins can occupy the same quantum energy level). The material is said to be *degenerate*. This means that the Boltzmann probability distribution [used in deriving Eq. (2.5) and related results] no longer applies and must be replaced by the Fermi–Dirac probability distribution. The two distributions are shown in Fig. 2.6 with the reference Fermi level (see Appendix E). They describe the probability of occupancy of a quantum state by an electron and the reference energy is called the *Fermi level*, E_F. This is the energy level at which there is a 50% probability of occupancy of a quantum state by an electron.

A very important property of the Fermi level is the fact that in thermal equilibrium (no current flowing, no incident light) it must be constant at every position. If this were not the case, a higher probabilty of occupancy would exist at one position compared to another and electrons would move, thus creating a current. At the end of Appendix F, the relation between E_F and doping level is derived. It is easy to see from equation (F.20a) that the Fermi level will be at the conduction-band edge when the electron concentration (which for N-type material is

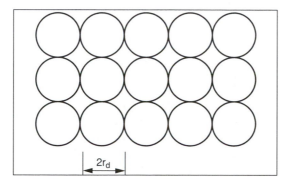

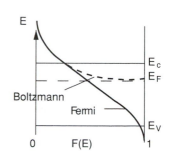

Figure 2.5 Orbitals of donor atom electrons for the critical concentration N_{crit}..

Figure 2.6 Fermi and Boltzmann probability distributions versus energy in a semiconductor.

practically the same as the donor concentration) is equal to $N_C \approx 10^{19}$ cm^{-3}, that is, comparable to the value obtained above. The net result of very heavy doping is a modification in some of our basic relations for heavily doped *N*-type semiconductors (where E_F lies in the conduction band, like a metal) and for heavily doped *P*-type materials (where E_F lies in the valence band). Degeneracy need not concern us in most of the work to be presented in this or the following chapters.

However, another even more important effect occurs, known as band-gap narrowing (BGN). At doping levels approaching N_{crit} the distribution of available quantum states in the conduction band extends downwards past the band edge E_c in a very complex manner, involving the quantum states close to the conduction band at the donor energy level and various particle–particle interactions. The net result is a reduction, or narrowing, in the value of the band gap energy E_g by an amount ΔE_g, as shown in Fig. 2.7. *This has a profound effect on all calculations involving minority carrier concentrations*, and can best be represented by considering an effective doping level N_{deff} to replace the actual doping level N_D, when considering minority carrier transport. The most widely used empirical expression for this effect is to write [4]:

$$\Delta E_g = E_{ref}\{\ln(N_D/N_{ref}) + [\ln(N_D/N_{ref})^2 + 0.5]^{0.5}\} \qquad (2.13)$$

where E_{ref} is of order 0.009 eV and N_{ref} is approximately 10^{17} cm^{-3} for typical cases. This is plotted in Fig. 2.8(a).

The corresponding value for effective doping level is:

$$N_{deff} = N_D \exp(-\Delta E_g/kT) \qquad (2.14)$$

This is plotted in Fig. 2.8(b). For a doping level $N_D = 10^{20}$ cm^{-3}, ΔE_g is of order 0.12 eV, and the effective doping level is approximately 10^{18} cm^{-3}. This effect will be shown to be very important when calculating minority carrier current flow and the current gain of bipolar transistors and should be taken into account when the actual doping level exceeds 10^{18} cm^{-3}.

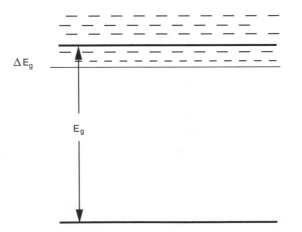

Figure 2.7 Penetration of occupied states by amount ΔE_g below E_c into the band gap.

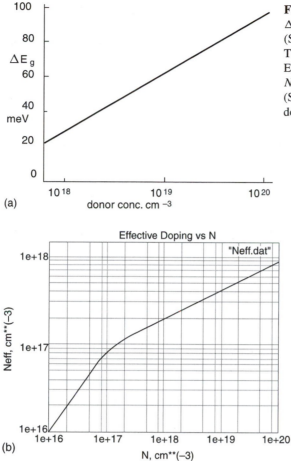

Figure 2.8 (a) Plot of band gap narrowing ΔE_g versus doping level N_D from Eq. (2.13). (See also CD-ROM BIPGRAPH example TRAN1; "Bandgap narrowing vs N.") (b) Effective doping N_{Deff} versus actual doping N_D for silicon at 300 K from BIPOLE3 [13]. (See also CD-ROM BIPGRAPH "Effective doping vs N.")

2.3 CURRENT FLOW

2.3.1 Drift current

Ohm's law is usually written as $V = IR$, where V is the voltage, I the current, and R the resistance:

$$I = V/R$$

Since $R = \rho L/A$, where ρ is the resistivity ($= 1/\sigma$, where σ is the conductivity), in terms of current density J we can hence write:

$$J = (V/L)A\sigma/A \tag{2.15}$$

which can be rewritten as:

$$J = E\sigma \tag{2.16}$$

In *N*-type material the conductivity is clearly proportional to the concentration of electrons available for conduction (i.e., the free-electron concentration n_n) and also to the charge on each electron q. The constant of proportionality is called the (electron) *mobility* μ_n, and the resulting expression for drift current density is:

$$J_n = q\mu_n n_n E \qquad (2.17)$$

The conductivity due to the electron concentration n_n is given by:

$$\sigma_n = q\mu_n n_n \qquad (2.18a)$$

Note that in general, with both an electron concentration n_n and a hole concentrtion p_n present, defining the hole mobility as μ_p, the conductivity is given by:

$$\sigma = q(\mu_n n_n + \mu_p p_n) \qquad (2.18b)$$

Since current is also proportional to the rate at which the electrons are moving (i.e., to their velocity), the electron current can also be written as:

$$J_n = qn_n v_d \qquad (2.19)$$

Hence the velocity is related to the mobility by $v_d = \mu_n E$. The current due to the electric field, E, is called a *drift current* (to distinguish it from another type of current to be dealt with later, called diffusion current) and v_d is referred to as the *drift velocity*. Even with no field applied, the electrons are moving under random thermal motion with collisions, as shown in Fig. 2.9 (in fact, the change in direction is due to interaction with the lattice or impurity atom). We can define a mean free time between collisions, t_m, and a mean free path length between collisions ℓ_m.

Consider the electron drifting with a velocity v_x due to an applied field E. The force acting on the electron of effective mass m_e is qE, also equal to $m_e a$, where a is the acceleration in the x direction between collisions. Hence we can write the acceleration in the x direction as:

$$a = qE/m \qquad (2.20)$$

The distance between collisions and hence the velocity acquired varies enormously in a random manner due to thermal motion. The normal relationship for the drift velocity acquired at the end of a time t_c (assuming the carrier starts with zero velocity after each collision) would be:

$$v_x = at_c \qquad (2.21)$$

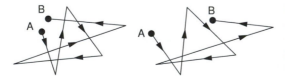

Figure 2.9 Illustration of random thermal motion of electron (a) in equilibrium, $E = 0$, where the distance $A - B \approx 0$; (b) in the presence of an applied electric field E, where the electron drifts a distance $A - B$.

The average x-direction velocity would thus be $at_c/2$. However, a detailed statistical derivation shows that if t_m is the *mean free time between collisions*, the result for the *average drift velocity* v_d is closely related to the result obtained from classical mechanics: $v_d = at_m$. Thus we have:

$$v_d = qEt_m/m \tag{2.22}$$

From our previous discussion, it follows that we can express the mobility as:

$$\mu = \frac{qt_m}{m} \tag{2.23}$$

or in terms of mean free path length between collisions:

$$\mu = \frac{q\ell_m}{mv_{th}} \tag{2.24}$$

where v_{th} is the thermal limited velocity of the electron due to its thermal energy. Since this energy is equal to $\frac{3}{2}kT$ and also equal to the kinetic energy $\frac{1}{2}mv_{th}^2$, v_{th} can be estimated at room temperature as:

$$
\begin{aligned}
v_{th} &= \sqrt{3kT/m} \\
&= \sqrt{3 \times 0.025 \times 1.6 \times 10^{-19}/9.11 \times 10^{-31}}\ \mathrm{m/s} \\
&= 1.14 \times 10^5\ \mathrm{m/s} \\
&= 1.14 \times 10^7\ \mathrm{cm/s}
\end{aligned}
\tag{2.25}
$$

Mobility dependence on electric field

The relationship between mobility and mean free time between collisions, t_m, assumed that the distance the carrier drifted in the x direction due to E_x was small compared to the total distance traveled in random directions between collisions. If the field E is large enough, this assumption is no longer valid. When the drift velocity becomes comparable to the thermal velocity, the mobility becomes strongly dependent on the value of E and eventually ceases to be a meaningful parameter. The velocity versus electric field saturates as shown in Fig. 2.10 for semiconductors like silicon and germanium. The special case of materials like GaAs, where a peak in the velocity versus field diagram occurs will be treated in Chapter 7.

The maximum or saturated drift velocity is an important value in semiconductor devices. Since the low-field mobility for moderately doped silicon is $\mu_n \approx 1000\ \mathrm{cm^2/V\ s}$ and at low

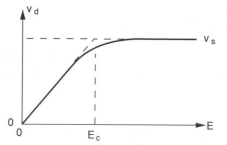

Figure 2.10 Drift velocity versus electric field for silicon and germanium.

TABLE 2.1
Values of E_c and v_s for silicon and germanium at moderate doping levels

	Si		Ge	
	electrons	holes	electrons	holes
E_c (kV/cm)	10	25	2	4
v_s (cm/s)	10^7	10^7	6×10^6	6×10^6

fields we can use $v_d = \mu_e E$, the value of critical field shown in Fig. 2.10 can be deduced as $E_c = 10^4$ V/cm. Note that the mobility is constant only for $E < 10^4$ V/cm. The other values for Si and Ge are given in Table 2.1. In nearly all semiconductor devices we will be concerned with both low- and high-field regions.

Mobility dependence on doping

The value of mobility depends on the mean free time between collisions. For light doping (below about 10^{16} cm^{-3}) the collisions occur mainly with the vibrating Si lattice atoms. As the impurity concentration (N_D or N_A) increases above about 10^{16} cm^{-3}, additional scattering occurs (the trajectory of the carrier (electron or hole) is altered by the coulomb force existing between the fixed charge on the ionized impurity atom and the carrier); the mobility starts to decrease. Figure 2.11 shows mobility versus doping curves for Ge, Si, and GaAs at room temperature.

These plots were generated by the BIPOLE3 simulation program [13]. (See also CD-ROM BIPGRAPH example TRAN1; "Electron and hole mobility vs *N*.")

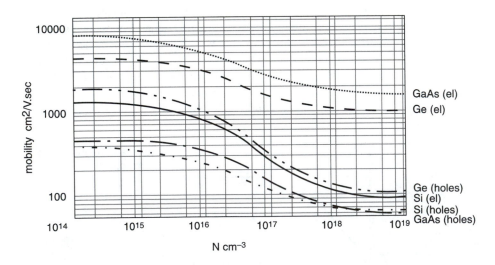

Figure 2.11 Mobility versus doping for electrons and holes in Ge, Si, GaAs at 300 K.

Mobility dependence on temperature

There are two main mechanisms determining the behavior of electron mobility versus temperature.

1. Low doping: lattice scattering dominates

In this case, as the temperature increases, the lattice atoms vibrate more about their mean position, thus effectively increasing in "size" and thereby increasing the chance of collisions. The mobility therefore decreases with increasing T. The theoretical dependence is $\mu \propto T^{-3/2}$.

2. High doping: ionized impurity scattering dominates

In this case, the dominating factor is the length of time for which the electron remains in the vicinity of the ionized impurity atom (the longer the time, the greater is the effect of the coulomb force). Since the electron is moving faster at higher temperatures, it will remain in the vicinity of the ionized impurity atom for a shorter time—the scattering effect is therefore decreased and the mobility increases with increasing temperature. The theoretical dependence is now $\mu \propto T^{3/2}$.

The net result is that because of the two competing mechanisms with opposite temperature coefficients, for moderate doping levels of order 10^{16} to 10^{17} cm^{-3}, the mobility varies only slowly with temperature. This is seen in Fig 2.12 for electrons and holes in silicon. At low doping levels the $T^{-3/2}$ law is approached. At high doping levels in silicon the $T^{+3/2}$ law is not normally observed at typical temperatures, but a distinct flattening of the curve nevertheless occurs due to the two opposing scattering mechanisms. In most design situations the values may be taken as constant with little loss of accuracy.

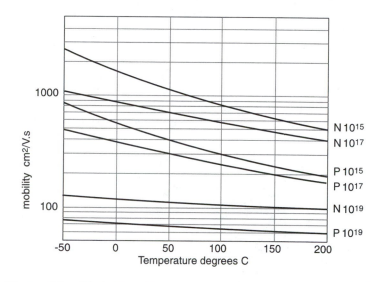

Figure 2.12 Electron (N) and hole (P) mobility versus temperature for silicon for doping levels of 10^{19}, 10^{17}, and 10^{15} cm^{-3}. These curves were generated using BIPOLE3 with mobility models developed from data in current literature.

2.3.2 Diffusion current

When an excess concentration of carriers (electrons or holes) is created at some point in a semiconductor (e.g., by shining light locally on one region, or by forward biasing a *PN* junction), the carriers tend to diffuse away from the point of high concentration to a region of low concentration because of their thermal energy giving rise to random thermal motion. This diffusion obeys Fick's law and is similar to smoke diffusing across a room [in the absence of a (drift) current of air]. The rate of flow is proportional to the concentration gradient. The current density is proportional to the rate of flow times the charge on one carrier. Consider electron flow as shown in Fig. 2.13:

$$J_n \propto q \frac{dn}{dx} \tag{2.26}$$

$$J_n = q D_n \frac{dn}{dx} \tag{2.27}$$

where we have introduced a constant of proportionality D_n.

This is called the diffusion constant or diffusion coefficient. Note that since electrons flowing in one direction create a current opposite to holes flowing in the same direction, and since conventional current flows in the opposite direction to electrons, the corresponding equation for hole diffusion current is:

$$J_p = -q D_p \frac{dp}{dx} \tag{2.28}$$

where D_p is the hole diffusion coefficient. The total current density in a semiconductor is given by the sum of diffusion and drift currents for both electrons and holes:

$$J_n = q D_n \frac{dn}{dx} + q \mu_n n E \tag{2.29}$$

$$J_p = -q D_p \frac{dp}{dx} + q \mu_p p E \tag{2.30}$$

Fortunately, in most cases involving hand analysis of semiconductor devices, only one of the four current terms will have to be considered in a particular case.

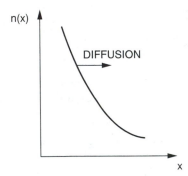

Figure 2.13 Nonconstant electron concentration versus distance to illustrate diffusion current.

The Einstein relation

Since both mobility (which determines drift current) and diffusion coefficient (which determines diffusion current) depend on random thermal motion of carriers, it is not too surprising that these two quantities are related. In Appendix G, we show that the ratio D_n/μ_n is a constant at a given temperature.

$$D_n/\mu_n = kT/q = V_t \tag{2.31}$$

This is called the Einstein relation, and it is clearly true for holes as well as for electrons. We have also introduced the quantity $V_t = kT/q$ (called the thermal voltage), which will be used frequently.

2.4 THE *P–N* JUNCTION

2.4.1 Basic properties

Figure 2.14 shows the carrier distribution and band diagram for a *P–N* junction under thermal equilibrium conditions (zero applied bias).

The electron concentration is high on the N side far from the junction (where it is equal to the doping level N_D), but on the P side it must be low, as given by the mass action law, and equal to n_i^2/N_A. For doping levels of 10^{16} cm^{-3} on each side, this gives, at room temperature, an electron concentration dropping from 10^{16} cm^{-3} on the N side to 10^4 cm^{-3} on the P side. There is clearly some region over which this transition must occur; furthermore, within this region there can no longer be space-charge neutrality, since the individual donor atoms on the N side always have their fixed positive ion charge, as discussed in Section 2.2. This region is referred to as a *space-charge region* (or a space-charge layer), a *transition region*, or a *depletion layer*.

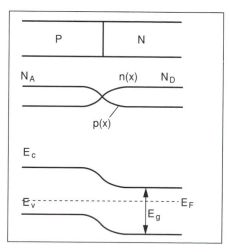

Figure 2.14 Free-carrier and band diagrams for a *P–N* junction.

It is a very important region of nearly all semiconductor devices and must be solved using Poisson's equation, which is another form of Gauss's Law.

$$\frac{dE}{dx} = (q/\epsilon)[N_D(x) - N_A(x) + p(x) - n(x)] \tag{2.32}$$

Here, and throughout the rest of this text, ϵ is defined as the dielectric constant, often written as the product $\epsilon_0\epsilon_r$, where these two constants are the permittivity of free space and the relative dielectric constant, respectively. The doping levels are represented as functions of distance x, as are the hole and electron concentrations. However, in all devices to be dealt with in this text, the doping levels are assumed to be constant on each side of a junction. Thus N_D is constant on the right and zero on the left of the junction in Fig. 2.14; the opposite comment applies to N_A. The change in potential resulting from the electric field is also reflected by a change in the conduction- and valence-band edges (the band gap E_g is a property of the material and remains constant), since under thermal equilibrium conditions the Fermi level must remain constant (otherwise current would flow). Equation (F.20) can be used to determine the position of E_F for any given doping level.

2.4.2 The Boltzmann relations for the *P–N* junction

We will now derive a basic relation between electron concentration and potential for a semiconductor in thermal equilibrium. It will be used frequently in our analysis of devices.

Consider a nonuniformly doped semiconductor in thermal equilibrium. The electron curent is zero:

$$J_n = 0 = qD_n \frac{dn}{dx} + q\mu_n n E \tag{2.33}$$

The electric field is thus given by:

$$E = -V_t \frac{1}{n} \frac{dn}{dx} \tag{2.34}$$

where $V_t = kT/q$. The potential between two points x_1 and x_2 (for example, the N and P sides in Fig. 2.14) is:

$$V = -\int E \, dx = V_t \ln(n) \tag{2.35}$$

Therefore:

$$V_{x_1} - V_{x_2} = V_t \ln \frac{n(x_1)}{n(x_2)} \tag{2.36}$$

Rearranging gives:

$$n(x_1) = n(x_2) \exp\{[V(x_1) - V(x_2)]/V_t\} \tag{2.37}$$

The final expressions, for both electrons and holes, are known as the *Boltzmann relations* (valid strictly only in thermal equilibrium):

$$n(x) = n_i e^{V(x)/V_t} \tag{2.38}$$
$$p(x) = n_i e^{-V(x)/V_t} \tag{2.39}$$

where for convenience we define $V(x) = 0$ at $x = 0$ where $n = p = n_i$.

2.4.3 Built-in barrier potential across a P–N junction

Consider any P–N junction with constant N- and P-type doping on either side. For electrons, the concentration on the N side is

$$n_{n0} = N_D \tag{2.40}$$

On the P side, the electron concentration is (using $np = n_i^2$, $p_{p0} = N_A$)

$$n_{p0} = n_i^2/N_A \tag{2.41}$$

Using the above Boltzmann relation for electrons gives the potential difference across the junction as:

$$V_{bi} = V_t \ln(n_{n0}/n_{p0}) \tag{2.42}$$

or

$$V_{bi} = V_t \ln \frac{N_A N_D}{n_i^2} \tag{2.43}$$

this is called the *built-in barrier potential* of the P–N junction. It means that even with no external voltage applied, there is an internal potential difference between the P and N sides of a P–N junction.

Substituting typical values for silicon: $N_A(P \text{ side}) = N_D(N \text{ side}) = 10^{16} \text{ cm}^{-3}$, $n_i = 1.6 \times 10^{10}$ cm^{-3}, $V_t = 0.0259$.
   This gives a value for $V_{bi} = 26.7 \times .0259 = 0.69V$.

Under *forward* bias, this barrier is reduced; under *reverse* bias it is increased.

2.4.4 Poisson's equation and the space-charge layer

Let us look more closely at the free-carrier distributions on Fig. 2.14 and see how the electron and hole concentrations together with the ionized donor and acceptor concentrations create a space charge on either side of the *P–N* junction. Figure 2.15 shows this region in detail. Diagram (b) shows the free-carrier concentration distributions $p(x)$ and $n(x)$. As already discussed, the hole concentration p must become practically equal to the acceptor concentration far from the junction on the left side, where space-charge neutrality exists; conversely the electron concentration is equal to the donor concentration N_D far on the right side. The minority carrier concentrations on either side are very small ($n_p = n_i^2/p_p = n_i^2/N_A = 10^6$ cm^{-3} on the *P* side, and similarly p_n on the *N* side has the same value in this example since it is a symmetrical *P–N* junction with doping levels of 10^{14} cm^{-3} on each side). It is clear that the electron and hole concentrations must become equal at some point and in this case of a symmetrical *P–N* junction the crossover point is at the junction, where $p = n = n_i$.

If we now consider the charge concentration $Q(x)$ (coulombs per cm^3) at any position x, it is given by:

$$Q(x) = q[(N_D(x) - N_A(x) + p(x) - n(x)] \tag{2.44}$$

(since each ionized donor atom contributes one positive fixed charge and each ionized acceptor atom contributes one fixed negative charge). The charge distribution must therefore have the appearance of diagram (c), becoming zero at some distance d_p from the junction on the *P* side and at some distance d_n from the junction on the *N* side.

The diagram for $dE/dx = Q(x)/\epsilon$ therefore has the general shape shown in diagram (c). Note that we have drawn diagram (b) on a logarithmic vertical scale and diagram (c) on a linear vertical scale for reasons that will become clear at a later stage. $Q(x)$ is practically constant for some distance on either side of the junction, since near the junction $p(x)$ and $n(x)$ each have values that are comparable to n_i, and if we add 10^{10} or even 10^{11} or 10^{12} electrons or holes per cm^3 to the impurity concentration 10^{14} cm^{-3}, the result is still close to 10^{14} cm^{-3}.

This enables us to make a very important simplification before solving Poisson's equation: *Providing the analysis is restricted to some (as yet undefined) region on either side of the junction, the free-carrier concentrations can be neglected in the calculation of electric field and potential distributions.* This is referred to as the *depletion approximation*, and the region under analysis is frequently referred to as the *depletion layer.* In Fig. 2.15 this extends from close to $-d_p$ on the *P* side to $+d_n$ on the *N* side. In fact, we shall assume that the space-charge layer terminates abruptly at $-d_p$ and $+d_n$ so that there is an abrupt transition from space charge to neutrality. Poisson's equation on the *N* side for $0 < x < d_n$ now becomes:

$$\frac{dE}{dx} = (q/\epsilon)N_D \tag{2.45}$$

Integrating once to obtain the electric field gives:

$$E(x) = (q/\epsilon)N_D(x - C_1) \tag{2.46}$$

where C_1 is a constant of integration. Since the electric field is normally very small in a neutral region (where $dE/dx = 0$), we set $E(d_n) = 0$, which gives the final result:

$$E(x) = (q/\epsilon)N_D(x - d_n) \tag{2.47}$$

The maximum magnitude of the electric field occurs at the junction ($x = 0$) and has the value:

$$E_{max} = (q/\epsilon)N_D d_n \tag{2.48}$$

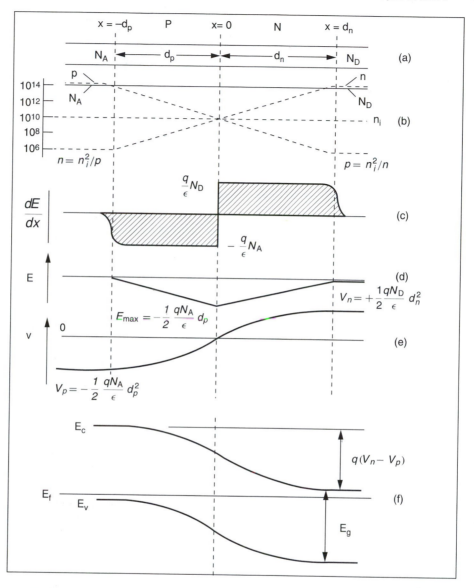

Figure 2.15 Symmetrical *P–N* junction diagrams for space-charge layer: (a) *P–N* junction; (b) free-carrier concentration distributions *p*, *n* (shown by broken lines); (c) space charge; (d) electric field; (e) potential; (f) band diagram. Reprinted with permission from D. J. Roulston *Bipolar Semiconductor Devices*, McGraw–Hill, New York, 1990.

Note that d_n is still not known.

Integrating a second time gives the potential distribution:

$$V(x) = -(q/\epsilon)N_D(x^2/2 - d_n x + C_2) \tag{2.49}$$

where C_2 is a new constant of integration. Since we are free to choose the reference for potential arbitrarily, we choose $V(x) = 0$ at $x = 0$, that is, at the junction. Thus C_2 is zero, and the final expression for potential on the N side is:

$$V(x) = -(q/\epsilon)N_D(x^2/2 - d_n x) \tag{2.50}$$

The total voltage drop on the N side will be of particular interest. This is the difference in potential between $x = 0$ and d_n and thus has the value:

$$V_n = (q/\epsilon)N_D(d_n^2/2) \tag{2.51}$$

By analogy, the total voltage drop on the P side is:

$$V_p = -(q/\epsilon)N_A(d_p^2/2) \tag{2.52}$$

Under zero-bias (thermal equilibrium) conditions, we have already seen that a built-in voltage V_{bi} exists across the junction. This must be equal to $V_n - V_p$. The zero-bias values for depletion layer widths can thus be obtained:

$$V_n - V_p = V_{bi} = 2(q/\epsilon)N_D(d_{nsym}^2/2) = (q/\epsilon)N_D d_{nsym}^2 \tag{2.53}$$

where we have introduced d_{nsym} for d_n to distinuish this result from that for a one-sided junction to be derived in the next section (2.61). Hence we have the required result for zero-bias depletion layer width on either side $d_{nsym} = d_{psym}$ for this symmetric $P–N$ junction:

$$d_{nsym} = d_{psym} = (\epsilon V_{bi}/q N_D)^{1/2} \tag{2.54}$$

It is instructive to get a feel for the magnitudes involved in typical semiconductor junctions. For the values used in Fig. 2.15, where each side is doped 10^{14} cm^{-3} (a very light doping level), the built-in barrier voltage, from Eq. (2.43), is (we use $kT/q = 0.0259$ V, corresponding to a temperature of 27°C, with $n_i = 1.6 \times 10^{10}$ cm^{-3}):

$$V_{bi} = 0.0259 \ln(10^{14} \times 10^{14}/2.56 \times 10^{20}) = 0.0259 \times 17.4 = 0.45 \text{ V}$$

The depletion layer thicknesses are therefore from Eq. (2.54) each equal to 1.7×10^{-4} cm $= 1.7 \, \mu$m.

Depletion layer under forward and reverse bias

It is a simple matter to extend the above results to the case where a bias voltage V_a is applied to the junction. Under forward bias this reduces the total potential drop across the $P–N$ junction from V_{bi} to $V_{bi} - V_a$. Likewise, under reverse bias, the potential across the junction becomes

$V_{bi} - V_a = V_{bi} + |V_a|$. The above results for peak electric field, total voltage, and depletion layer thickness may thus be used, replacing V_{bi} by $V_{bi} - V_a$.

In the above numerical example, at a forward bias of $V_a = 0.35$ V, the depletion layer widths decrease from 1.7 μm to $1.7[(0.45 - 0.35)/0.45]^{1/2} = 0.83$ μm.
For a reverse bias of 10 V, the value of $d_p = d_n$ is $1.7(10.45/0.45)^{1/2} = 8.1$ μm.

In general it can be assumed that under forward bias the depletion layer thickness is small compared to the neutral region width in a semiconductor device; this assumption cannot be made in reverse bias.

2.4.5 The "one-sided" or asymmetric abrupt *P+–N* junction

Consider now a *P–N* junction in which the *P* side is doped much more heavily than the *N* side [N_A (*P* side) $\gg N_D$ (*N* side)]. From charge-balance considerations, the total charge Q_n, Q_p on each side must be the same magnitude (although of opposite sign), as shown in Fig. 2.16:

$$Q_n = qAN_Dd_n = qAN_Ad_p = Q_p \tag{2.55}$$

hence:

$$N_Dd_n = N_Ad_P \tag{2.56}$$

For N_A (*p* side) $\gg N_D$ (*n* side) we thus have:

$$d_n/d_p \gg 1$$

Using the expressions for voltage already derived in Eqs. (2.51) and (2.52), the ratio of the voltage drops on either side of the junction V_n/V_p is:

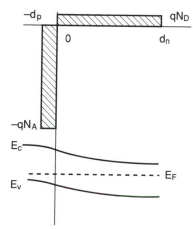

Figure 2.16 The *P+–N* nonsymmetric junction showing space-charge regions (shaded) and energy-band diagram.

$$V_n/V_p = N_D d_n^2/(N_A d_p^2) \tag{2.57}$$

or

$$V_n/V_p = (N_D d_n/N_A d_p)(d_n/d_p) \tag{2.58}$$

Substituting from Eq. (2.56) gives:

$$V_n/V_p = d_n/d_p \tag{2.59}$$

For the *P+–N* junction, this is therefore much greater than unity. We conclude that for a one-sided abrupt *P–N* junction, the depletion layer thickness and the potential on the heavily doped side may be neglected in calculations. Hence for this asymmetric *P+–N* junction:

$$E_{\text{max}} = (q/\epsilon)N_D d_n \tag{2.60}$$

$$V_{\text{tot}} = (q/2\epsilon)N_D d_n^2 \tag{2.61}$$

This is a very common situation in real semiconductor devices, since junctions are often made by diffusing a high concentration of, for example, phosphorus atoms, into a background of moderately (boron) doped material (*N+–P* junction) or vice versa (*P+–N* junction). This result can be rearranged to give the depletion layer width, which is proportional to the square root of the total voltage.

$$d_n = \sqrt{2\epsilon V_{\text{tot}}/q N_D} \tag{2.62}$$

For *N+–P* junctions in which the doping is high on the *N* side, the above expressions may be used with N_D replaced by N_A and d_n replaced by d_p. Note that it is the doping on the lightly doped side that enters into all depletion layer calculations.

The energy-band diagram is also shown in Fig. 2.16. Note that the Fermi levels align throughout the device in thermal equilibrium, and the potential follows the conduction-band edge with a sign change $[E(x) = -qV(x)]$ with the Fermi level position in the *P* and *N* material being given by Eqs. (F. 20a) and (F.20b).

Transition between depletion layer and neutral region

In the above treatment [in writing Eq. (2.45)] we assumed an abrupt change from depletion to neutrality; in Fig. 2.16 the space charge is shown as dropping abruptly from a value qN_D to zero at a distance d_n from the junction. In reality there is a gradual transition, as shown in Fig. 2.15(c), occurring over a distance of a few *extrinsic Debye lengths*. The Debye length is a function of doping level, *N*, given by:

$$L_D = (kT\epsilon/q^2 N)^{1/2} \tag{2.63}$$

In most cases this is considerably shorter than d_n or d_p and the transition from complete depletion to neutrality may be assumed to be abrupt with little loss of accuracy.

Consider a junction with a *P*-side doping level $N_A = 10^{19}$ cm^{-3} and an *N* side doping level $N_D = 10^{16}$ cm^{-3}. The builtin barrier potential is:

$$V_{bi} = 0.025 \ln \frac{10^{19} \times 10^{16}}{10^{20}} = 0.86 \text{ V}$$

The depletion layer thickness is from Eq. (2.62)

$$d_n = 0.32 \, \mu\text{m at zero bias } (V_{tot} = 0.86)$$

If a reverse bias of 10 V is applied, the depletion layer thickness becomes:

$$d_n = 0.32 \, \mu\text{m}\sqrt{10.86/0.86} = 1.2 \, \mu\text{m}$$

The Debye length from Eq. (2.63) is 0.04 μm, small enough to be neglected under reverse bias, but not quite negligible at zero bias.

2.4.6 The linearly graded junction

In diffused junctions like Fig. 1.14(a), and for narrow depletion layers (typically around zero volts and forward bias), a good approximation to the doping profile is a linear gradient as shown in Fig. 2.17.

$$N(x) = ax \tag{2.64}$$

If this is used in Poisson's equation (2.45) and integrated once, using the condition that the field is zero at the depletion layer boundaries, one obtains:

$$\frac{dV}{dx} = -(aq/\epsilon)(x^2 - d_n^2) \tag{2.65}$$

Integrating a second time and defining $V = 0$ at $x = 0$ as in Eq. (2.50) gives:

$$V(x) = -(aq/\epsilon)(x^3/3 - d_n^2 x) \tag{2.66}$$

The total voltage $V_{n\text{lin}}$ on the *N* side is thus:

$$V_{n\text{lin}} = -(aq/\epsilon)(2/3)d_n^3 \tag{2.67}$$

This may be compared to Eq. (2.52), where the total voltage was proportional to d_n^2. The result will be used when discussing depletion layer capacitance laws.

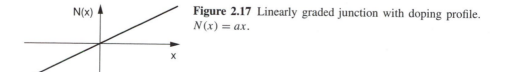

Figure 2.17 Linearly graded junction with doping profile. $N(x) = ax$.

2.4.7 Minority carrier injection under forward bias

We will now consider a very important situation. Let the junction be forward biased by some small or moderate voltage (a few tenths of a volt less than the built-in barrier potential V_{bi}). Figure 2.18 represents this situation for an *N–P* junction.

Let us consider electron current flow given by Eq. (2.29):

$$J_n = qD_n\frac{dn}{dx} + q\mu_n nE$$

It is clear that the "average" gradient is very large across this space-charge region, since n varies from typically 10^{18} on the *N* side to a value many orders of magnitude smaller on the *P* side, over a distance less than 1 μm. The diffusion current is thus very high. Likewise, the electric field is high, of order $(V_{bi} - V_a)/w$, where w is the total depletion layer width $(d_n + d_p)$. If the current is zero, these two terms must cancel; this was the condition used in deriving the Boltzmann relations. However, it is clear that the current need not be exactly zero; as long as J_n is much less than either the diffusion or drift components, it is a very good approximation to say that these two components still (almost) cancel each other within the space-charge layer (but certainly not outside this region, where the electric field is very small and the carrier concentration gradient is also small).

We can therefore write, *inside the space-charge layer to a good approximation* under small or moderate forward bias:

$$qD_n\frac{dn}{dx} + q\mu_n nE = 0 \qquad (2.68)$$

Using exactly the same procedure as for the derivation of the Boltzmann relation for electrons, using the Einstein relation to substitute $D_n/\mu_n = kT/q = V_t$, this leads to:

$$V_{jtot} = V_t \ln[n_n(0)/n_p(w)] \qquad (2.69)$$

where V_{jtot} is the total potential across the space-charge layer width w, $n_n(0)$ is the electron concentration on the *N* side, $n_p(w)$ is the electron concentration on the *P* side. On the *N* side, the electron concentration is practically equal to the donor concentration [it is in fact equal to $N_D + p_n(0)$, but $p_n(0)$ is orders of magnitude smaller than N_D even for moderate forward bias]. Substituting N_D for $n_n(0)$ and rearranging gives:

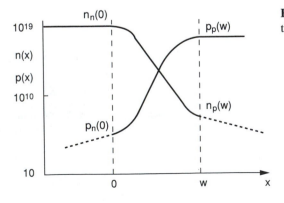

Figure 2.18 Electron and hole distributions in the space-charge layer under forward bias.

$$n_p(w) = N_D \exp(-V_{jtot}/V_t) \qquad (2.70)$$

Now the total potential across the junction under forward bias is

$$V_{jtot} = V_{bi} - V_a \qquad (2.71)$$

Hence we obtain:

$$n_p(w) = N_D \exp(-V_{bi}/V_t) \exp(V_a/V_t) \qquad (2.72)$$

Substituting for $V_{bi} = V_t \ln(N_A N_D/n_i^2)$ gives:

$$n_p(w) = (n_i^2/N_A) \exp(V_a/V_t) \qquad (2.73)$$

Since n_i^2/N_A is the thermal equilibrium value of the electron concentration on the *P* side, this is usually written as:

$$n_p(w) = n_{p0} \exp(V_a/V_t) \qquad (2.74)$$

with the derivation for holes being analagous and need not be derived separately:

$$p_n(0) = p_{n0} \exp(V_a/V_t) \qquad (2.75)$$

These two equations are called the *law of injection* for minority carriers across a *P–N* junction.

To see typical values, consider $N_D = 10^{16}$ cm^{-3} on the *N* side, $N_A = 10^{16}$ cm^{-3} on the *P* side.

The thermal equilibrium electron and hole concentrations (n_i^2/N) are thus each approximately equal to $2 \times 10^{20}/10^{16} = 2 \times 10^4$ cm^{-3} (assuming $n_i^2 = 2 \times 10^{20}$).

For $V_a = 0.5$ V, using $V_t = 0.026$ V, $\exp(V_a/V_t) = 2.25 \times 10^8$, gives $n_p(w) = p_n(0) = 4.5 \times 10^{12}$ cm^{-3}. This is increased by approximately a factor of 50 times for each 0.1 V increase in V_a.

Note that using slightly different values for n_i or V_t (or slightly different temperatures) will change the values of $n_p(w)$ and $p_n(0)$ significantly.

When dealing with carrier flow in the neutral regions (outside the space-charge layer), the depletion layer thickness, w, is usually much smaller than the thickness of the neutral region and the depletion layer boundaries are very close to the junction; the reference position for injected carriers in dealing with the neutral regions is thus taken to be very close to the junction.

Note on the validity of the injection law

To test the validity of the assumptions under which this injection law was derived, let us substitute some typical numbers for the quantities in the diffusion and drift current terms.

Based on the preceding discussion for the P^+–N junction, with $w = 0.3\,\mu$m at zero bias; the "average" gradient is of order $10^{19}/0.3 \times 10^{-4}$, giving $J_{n(\text{diffusion})} = q D_n (dn/dx)$ of order 10^6 A/cm^2. Alternatively, from the total potential V_{bi} and the depletion layer thickness, we estimate the "average" electric field to be of order 2×10^4 V/cm, thus giving an "average" electron drift current density $q \mu_n n E = 3 \times 10^{-12} n$ A/cm^2; since n starts at 10^{19} cm^{-3} on the N side in the above example, it is clear that at least over a substantial part of the space-charge layer, the individual electron drift and diffusion currents are extremely high. The distribution of electron concentration is very nonlinear and we cannot proceed any further with these estimates. It is quite clear, however, that the terminal current density J_n can certainly exceed several hundred amperes per cm^2 before the assumptions (of drift and diffusion current cancelling each other) become invalid. Note also that from Eqs. (2.69) and (2.71), when the applied bias across the space-charge layer is increased to a value V_{bi}, $n_n(0) = n_p(w)$ and likewise $p_n(0) = p_p(w)$. In this case the space charge has "collapsed" and the potential barrier is reduced to zero. In fact, the assumptions used in the above analysis are not valid at this high forward bias, but the conclusion can still be drawn that the barrier is reduced to zero for high forward bias.

2.5 NEUTRAL REGION RECOMBINATION AND THE CONTINUITY EQUATION

2.5.1 Recombination

Recombination is the process whereby excess electrons in the conduction band tend to decay towards the thermal equilibrium value by recombining with holes in the valence band. There are several possible scenarios. The first is band-to-band or direct recombination with emission of photons of energy E_g. This is shown diagrammatically in Fig. 2.19(a) and occurs in materials such as GaAs (referred to as a direct-gap semiconductor) but not in silicon.

Band-to-band recombination can also occur without photon radiation, the excess energy being given instead to an electron in the conduction band and thence to the crystal as thermal energy; this is the Auger recombination process.

Third (the most common mechanism in silicon), recombination occurs through intermediate levels called recombination centers (between E_C and E_V), as shown in simplified form in Fig. 2.19(b). Recombination centers near the middle of the band gap will have the greatest influence in increasing the chance of a recombination event, thereby decreasing the carrier lifetime. The centers are due either to imperfections in the crystal lattice, or to the introduction

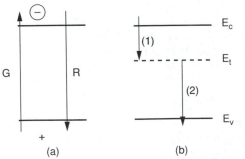

Figure 2.19 (a) Illustration of generation and recombination across the band gap; (b) indirect recombination via recombination centers at energy level E_t.

of elements such gold atoms, which can be used to control the carrier lifetime. Referring to Fig. 2.19(b), the electron is captured by a recombination center at energy level E_t and then subsequently recombines with a hole in the valence band (the same as capture of a hole from the valence band). The more recombination centers, the lower is the carrier lifetime τ.

For electrons the recombination rate R is the rate at which excess electrons, n', recombine:

$$R = \frac{dn'}{dt} \tag{2.76}$$

R is proportional to the number of electrons in the conduction band and to the number of holes in the valence band. Recombination is characterized by the carrier lifetime τ. This is often given a subscript so that τ_n is the electron lifetime in P-type material and τ_p the hole lifetime in N-type material. However, it should be recognized that the electrons and holes always recombine with each other so that in neutral regions of a semiconductor the excess carrier concentrations are the same. Furthermore, the carrier lifetime is strongly doping dependent; so the actual values of τ_n and τ_p in different regions vary with position. In general, we can write:

$$R = \frac{dn'}{dt} = \frac{n'}{\tau} \tag{2.77}$$

For the case of uniform light illumination creating an excess electron concentration n' with the light source turned off at time $t = 0$, the solution of Eq. (2.77) is:

$$n'(t) = n'(t = 0) \exp(-t/\tau) \tag{2.78}$$

as stated in discussing Fig. 1.11.

Carrier lifetime is very dependent on material preparation and device processing. Values vary from less than 1 ns for highly doped silicon (doping levels of order 10^{21} cm^{-3}), to more than 10 μs in lightly doped silicon (doping levels less than 10^{15} cm^{-3}).

Surface recombination

At the surface of a semiconductor the crystal structure is terminated. This results in a large number of recombination centers being introduced. In order to characterize the way in which excess carriers recombine in this situation, a parameter S called *surface recombination velocity* is used. It defines the relation between excess carriers and current density. For electrons the relationship is given by:

$$J_n = q S_n n' \tag{2.79}$$

Surface recombination velocity may in fact be used to characterize any surface; for example, an N–N^+ junction, a polysilicon–silicon interface, a metal semiconductor interface, an Si–SiO$_2$ interface. Typical values range from 10^6 cm/s for a metal contact to 1 cm/s for a silicon to oxide interface.

2.5.2 Diffusion in the presence of recombination: the continuity equation

Let us now consider what happens in the neutral region of a P–N junction when carriers are diffusing in the presence of recombination. Figure 2.20 shows a carrier distribution, with electrons diffusing from left to right.

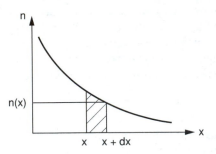

Figure 2.20 Electrons diffusing in the presence of recombination.

Recombination of electrons with holes implies that the electron current I_n is converted to hole current I_p (it is clear that the total current $I_p + I_n$ must remain constant).

The electron current $I_n(x)$ entering the shaded area at x is reduced by an amount dI_n within the interval dx. The incremental loss dI_n is due to the excess electron charge Q_n in the interval dx recombining with a lifetime τ_n. We can thus write:

$$I_n(x + dx) - I_n(x) = dI_n = Q_n/\tau_n \tag{2.80}$$

The volume of charge is simply:

$$Q_n = q A n'(x) \, dx \tag{2.81}$$

where we use $n'(x)$ to represent excess electron concentration $n(x) - n_{p0}$. Combining these two equations gives:

$$\frac{dI_n}{dx} = q A n'(x)/\tau_n \tag{2.82}$$

Substituting the expression for electron diffusion current, $I_n = q A D_n \, dn/dx$ gives:

$$\frac{d^2n}{dx^2} = n'(x)/(D_n \tau_n) \tag{2.83}$$

This is usually written as:

$$\frac{d^2n}{dx^2} = n'(x)/L_n^2 \tag{2.84}$$

where

$$L_n = \sqrt{(D_n \tau_n)} \tag{2.85}$$

and is called the *electron diffusion length*. It is clear that if the distance under consideration is comparable to or greater than a diffusion length, recombination will be important (i.e., the current of a given type will vary with distance).

Clearly, we could have performed the same calculations for holes instead of electrons recombining. This would have yielded:

$$\frac{d^2 p}{dx^2} = p'(x)/L_p^2 \tag{2.86}$$

where $L_p = \sqrt{(D_p \tau_p)}$ is the hole diffusion length for a hole lifetime τ_p.

Charge continuity equation

A particular case of the above result is when the electron current has decayed to zero at the right-hand side of the region under consideration. In this case we can write the current at the left side as:

$$I_n(0) - I_n(\text{rhs}) = I_n(0) = Q_{ntot}/\tau_n \tag{2.87}$$

where Q_{ntot} is the total electron charge integrated over the whole region. This is referred to as the *charge continuity equation* and will be of use in specific cases. A similar result applies to holes.

2.6 CONCLUSIONS

In this chapter we saw that a semiconductor can be characterized essentially by its band gap E_g and that this quantity can be determined from measurements of electrical conductivity over a suitable range of temperature. We also saw that there are two types of carrier—electrons and holes—and that the dominant type is determined by the dopant species (donor or acceptor). The mass action law $np = n_i^2$ was shown to relate the electron and hole concentrations in thermal equilibrium, with n_i also being available from conductivity versus temperature measurements. Carrier mobility μ and drift velocity v_d were introduced to describe current flow due to an applied electric field, and the concept of saturated drift velocity v_s was explained and the value estimated. Diffusion was presented as another mechanism for carrier flow, and the Einstein relation was stated to relate diffusion coefficient and mobility.

The *P–N* junction was introduced, with the Boltzmann equations relating electron and hole concentration to potential in thermal equilibrium. The concept of built-in barrier potential was described, and Poisson's equation was solved for the space-charge layer (depletion layer) on either side of a junction. The *P+–N* junction was examined and the injection law derived. The chapter concluded with a brief introduction to recombination and the continuity equation, which is of importance primarily in high-voltage diodes and thyristors. The use of materials with different band gaps in a single device to form heterojunctions will be dealt with subsequently (Chapter 7).

These results will be used in subsequent chapters to obtain the electrical terminal characteristics of diodes and transistors.

PROBLEMS

1. Using Eq. (2.2) and assuming that the only material parameter change is due to the band gap energy, calculate the intrinsic carrier concentration in the following materials (use Table 1.1 for values of E_g): (a) Carbon; (b) GaAs ; (c) Si; (d) Ge.

2. Calculate the hole concentration in silicon doped with phosphorus concentrations of: (a) 10^{21}; (b) 10^{18}; (c) 10^{16}; (d) 10^{14} cm^{-3}.

3. Calculate the percentage error in using Eq. (2.7) for calculating the minority carrier concentration p_{n0} in silicon at room temperature, doped $N_D = 10^{14}$ cm^{-3}.

4. Assuming a mobility of 1000 cm^2/V s, calculate the resistivities for the four cases considered in Problem 2. Repeat the calculations using mobility values extracted from Fig. 2.11.

5. Calculate the ratio of the contributions of holes to electrons in determining the conductivity of *N*-type silicon doped 10^{13} and 10^{16} cm^{-3} at room temperature. Repeat for *P*-type silicon for the same two doping levels.

6. Calculate the temperature at which the intrinsic carrier concentration n_i in silicon becomes equal to the background doping N if $N = 10^{16}$ cm^{-3}.

7. Calculate using Eq. (E.2) the error in using the Boltzmann distribution compared to the Fermi–Dirac distribution at the conduction-band edge, for *N*-type silicon doped: (a) 10^{16}; (b) 10^{17}; (c) 10^{18} cm^{-3}.

8. Use Eqs. (2.13) and (2.14) to calculate and plot (using log–log scales) the effective doping level N_{deff} in silicon over the doping range 10^{16} to 10^{21} cm^{-3}.

9. Use Eq. (2.24) and Fig. 2.11 to calculate for electrons the mean free path length between collisions for silicon doped: (a) 10^{14}; (b) 10^{17}; (c) 10^{19} cm^{-3}.

10. Use an electron mobility value for silicon equal to 1000 cm^2/V s plus Eqs. (2.23) and (2.24) to calculate the mean free time between collisions and the mean free path length between collisions. Compare the latter quantity to the interatomic spacing in silicon and comment on the frequency of collisions. Calculate the distance between donor atoms if the doping level is 10^{16} cm^{-3} and comment on the probable effect of ionized impurity atom scattering at this and at higher doping levels.

11. Calculate the electron current density that flows in silicon doped $N_D = 10^{16}$ cm^{-3} when an electric field of 1 V/cm is applied. Hence determine the hole concentration gradient required to give this magnitude of *diffusion* current. If the hole concentration $p(x)$ of diffusing holes is created by injection across a junction and decays linearly with distance, at what distance will $p(x)$ have reached zero if $p(0) = N_D$?

12. Calculate the built-in barrier potential for the following *P–N* junctions: (a) $N_A = N_D = 10^{15}$ cm^{-3}; (b) $N_A = N_D = 10^{17}$ cm^{-3}; (c) $N_A = N_D = 10^{19}$ cm^{-3}; (d) $N_A = N_D = 10^{21}$ cm^{-3}; (e) $N_A = 10^{19}$ cm^{-3}; $N_D = 10^{16}$ cm^{-3}. Calculate the total depletion layer thickness for each case.

13. For the junctions of Problem 12 calculate the depletion layer thickness for a reverse bias of 10 V. Could this bias be applied in all cases? *Hint:* Calculate the peak value of electric field in each case.

14. Use the injection law for a diode with $N_A = 10^{19}$, $N_D = 10^{16}$ cm^{-3} to calculate the forward voltage required to raise the hole concentration on the *N* side to a value equal to the donor concentration. Calculate the corresponding value of electron concentration injected on the *P* side. Would you expect the injecton law to remain valid at higher forward-bias values?

15. The resistivity ρ of a semiconductor sample is measured at different temperatures, with the following results: 50°C: 700 ohm cm; 100°C: 1000 ohm cm; 150°C: 230 ohm cm; 200°C: 33 ohm cm. Sketch this curve on a $\log(\sigma)$ versus $(1/T_K)$ plot and then determine the band gap energy E_g. Hence estimate the doping level of the sample.

16. Light from a GaAs light-emitting diode falls on one end of a silicon bar perpendicular to the length in the x direction at $x = 0$ (i.e., no light is absorbed at $x > 0$). If the lifetime of the carriers is 1 μs, calculate: (a) the fractional decrease in carrier concentration at a distance 100 μm from $x = 0$; (b) the fractional decrease in carrier concentration at a time 10 μs after turning off the light.

17. A P^+–N silicon diode has $N_A = 10^{18}$ cm^{-3}, $N_D = 10^{16}$ cm^{-3}. Sketch (with values) the carrier concentration diagram (on a log–linear plot) within the space-charge layer at zero bias. Calculate the value of applied bias across the depletion layer at which the barrier would be reduced to zero; sketch $n(x)$ and $p(x)$ for this case. Estimate the "average" hole diffusion current density within the space-charge layer for a bias of: (a) 0; (b) 0.3; (c) 0.6 V. Hence estimate very approximately the maximum terminal current density in each case for which you might expect the injection law to be valid to about 1% accuracy.

18. A P–N junction is doped 10^{16} cm^{-3} on each side. Calculate the depletion layer widths and the injected carrier concentrations at room temperature for applied forward bias equal to: (a) 0.5; (b) 0.6; (c) 0.65; (d) 0.7 V. Sketch $p(x)$ and $n(x)$ inside the depletion layer using a linear vertical scale. Hence sketch the space-charge diagram and comment on the meaning of "depletion layer" for high forward bias.

References

1. W. C. Dunlap, *An Introduction to Semiconductors*. New York: Wiley, 1975.
2. J. C. Moll, *Physics of Semiconductors*. New York: McGraw–Hill, 1964.
3. S. M. Sze, *Physics of Semiconductor Devices*. 2nd ed., New York: Wiley-Interscience, 1981.
4. J. W. Slotboom and H. C. de Graaf, "Measurements of bandgap narrowing in Si bipolar transistors," *Solid State Electronics* **19**, 857–62 (1976).
5. E. S. Yang, *Fundamentals of Semiconductor Devices*. New York: McGraw–Hill, 1988.
6. B. G. Streetman, *Solid State Electronic Devices*. Englewood Cliffs, New Jersey: Prentice Hall, 1990.
7. A. Bar-Lev, *Semiconductors and Electronic Devices*. 3rd Ed., New York: Prentice-Hall, 1993.
8. D. H. Navon, *Semiconductor Microdevices and Materials*. New York: Holt, Rinehart, Winston, 1986.
9. M. S. Tyagi, *Introduction to Semiconductor Materials and Devices*. New York: Wiley, 1991.
10. R. B. Adler, A. C. Smith, R. L. Longini, *Introduction to Semiconductor Physics*. Semiconductor Electronics Education Committee, Vol. 1, New York: Wiley, 1964.
11. P. E. Gray, D. DeWitt, A. R. Boothroyd, J. F. Gibbons, *Physical Electronics and Circuit Models of Transistors*. Semiconductor Electronics Education Committee, Vol. 2, New York: Wiley, 1964.
12. D. L. Pulfrey and N. G. Tarr, *Introduction to Microelectronic Devices*. Englewood Cliffs, New Jersey: Prentice-Hall, 1989.
13. D. J. Roulston, *BIPOLE3 Users Manual*. BIPSIM, Inc., and University of Waterloo, June 1996
14. C. Kittel, *Introduction to Solid State Physics*. New York: John Wiley, 1976.
15. D. A. Neamen, *Semiconductor Physics and Devices*. Chicago: Irwin, 1997.

Chapter 3

P–N Junction Diodes

3.1 INTRODUCTION

In this chapter we examine how current flows in *P–N* junction diodes. There are several types of diode in use today, but we will limit our discussions to those types that are either fundamental in demonstrating basic theoretical properties or have substantial practical importance. The *narrow-base diode* demonstrates some very basic behavior concerning current flow and is also exceedingly simple to examine theoretically. Furthermore, it forms the basis of the most critical part of the bipolar junction transistor to be dealt with in the following chapter. The results apply to both discrete and integrated BJT structures. This will therefore be our starting point. After deriving the *I–V* characteristics for this structure, we examine the *PIN diode* rectifier, the most widely used discrete diode structure; this also forms an introduction to *high-level injection* effects. Only the ideal *PIN* diode is considered; this keeps the treatment simple and yet demonstrates some of the crucial high-current effects to be seen in a wide range of devices.

A short overview will be presented of the wide-base diode (of little practical importance, but which demonstrates some useful properties) and the *P+NN+* diode (a special case of the *PIN* diode). Low forward and reverse bias effects due to space-charge recombination and generation are then studied. This treatment applies to all diode types and concludes with an overview of *avalanche multiplication* and *breakdown voltage*. Capacitive effects due to the depletion layer and to minority carrier charge storage are then studied, leading to the small-signal equivalent circuit of the diode.

Switching transients in diodes are explained. This will be sufficient to introduce the student to the concept of free-carrier charge injection and evacuation, leading to some very simple results for estimating switching times. The *temperature dependence* of diode *I–V* characteristics is analyzed, and the chapter concludes with a brief treatment of *metal–semiconductor contacts* and the *Schottky diode*.

3.2 THE *P+–N* NARROW-BASE DIODE

Figure 3.1 shows the impurity atom distribution for an ideal *P+–N* narrow-base diode. The *P*-region doping level N_A is assumed to be high, of order 10^{19} cm^{-3}, compared to the N_D-region doping level (in the range 10^{14} to 10^{17} cm^{-3}).

The thickness of the N and P layers is small enough that no current is lost through recombination; that is, W_n and W_p are each considerably less than the respective electron and hole diffusion lengths. Ohmic contacts are assumed at distances W_p, W_n from the junction on either side. A property of a good ohmic contact is that all carriers that reach the plane of the contact recombine at that plane. Therefore, whatever the value of the injected electron and hole concentrations at the junction, the excess carrier concentrations are forced to be zero at the edges of the neutral regions. In the bipolar transistor, the metal contact at W_n is replaced by a reverse-biased junction; this has an almost identical effect on the minority carriers, which is why the results to be obtained are so important.

In practice the diode is made by diffusing a P layer (normally boron) into a moderately doped N region, as shown in Fig. 3.2. The diffused P region does not have a constant doping profile. However, as we shall shortly see, this does not seriously affect the result of our analysis, since it is always the lightly doped side that determines the most important terminal electrical characteristics of the diode. The lightly doped region is assumed to be thin in the following analysis. In practice this will often be true only for the case where the metal is replaced by the reverse-biased junction of a BJT, but it is a convenient starting point and can be realized in practice by having a heavily doped N^+ region region with a low lifetime (or an interface between the N and N^+ regions with a low lifetime, as can occur easily during manufacture of an N epitaxial layer (see Section 8.2.2) on a wafer made from an N^+ substrate).

Let us now consider the electron and hole distributions when a forward bias V_a is applied to the diode. We use the notation $x = 0$ on the N side to be the depletion layer boundary on that side and likewise for the P side. From the injection law derived in Chapter 2, Eq. (2.74), we know that the electron concentration $n_p(0)$ on the P side of the depletion layer is raised to a value:

$$n_p(0) = n_{p0} \exp(V_a/V_t) \tag{3.1}$$

where $n_{p0} = n_i^2/N_A$ is the thermal equilibrium electron concentration on the P side. Likewise on the N side of the depletion layer, from Eq. (2.75), the injected hole concentration is raised to a value:

$$p_n(0) = p_{n0} \exp(V_a/V_t) \tag{3.2}$$

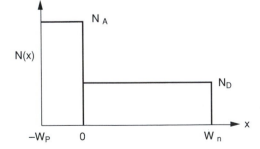

Figure 3.1 Impurity doping distribution for a P^+–N narrow-base diode.

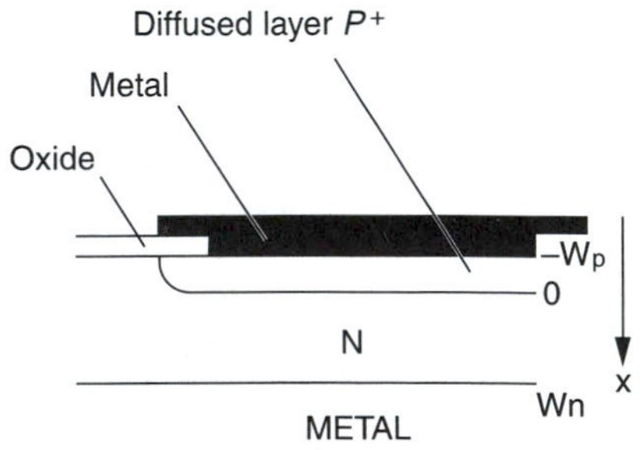

Figure 3.2 P^+–N narrow-base diode: practical realization.

Since we know that the ohmic contacts have the property of allowing no excess carrier concentrations at the metal–semiconductor interface, and since we have already said that the neutral region widths are sufficiently small that recombination can be neglected, it follows that the electron distribution may be assumed to decrease linearly from the value $n_p(0)$ at the junction to the value n_{p0} at the metal contact a distance W_p from the junction. Assuming that the total depletion layer thickness is much less than the neutral region thicknesses, the free carrier distribution must therefore be as shown in Fig. 3.3.

The vertical axis is shown broken so that the majority carrier concentration distributions can be visualized. From charge neutrality considerations, in the N neutral region the sum of the charge concentrations must add up to zero; hence

$$n_n(x) = N_D + p_n(x) \tag{3.3}$$

and on the P side:

$$p_p(x) = N_A + n_p(x) \tag{3.4}$$

The majority carrier distributions are thus identical in shape to the minority carrier distributions but raised above the respective (constant) doping levels. Since the relative change in majority carrier concentrations is very small [e.g., if $p(0) = 10^{13}$ cm^{-3} and $N_D = 10^{16}$ cm^{-3}, $n_p(0) = 1.001 N_D$, an increase of only 0.1%], it can be assumed that the current will be determined entirely by the change in *minority carrier* distributions. This will be evident for a P^+N diode, where the electron current injected into the P^+ region will be seen to be very small compared to the hole current injected into the N region. Furthermore, there is no reason for any substantial minority carrier drift current to flow, since the electric field in the neutral region is small. This argument is not rigorous, but the conclusion is in fact correct, as is shown in Appendix H. The hole current is therefore due almost entirely to diffusion:

$$J_p = -q D_p \frac{dp}{dx} \tag{3.5}$$

$$= q D_p [p_n(0) - p_{n0}] / W_n \tag{3.6}$$

Substituting from the injecton law for $p_n(0) = p_{n0} \exp(V_a/V_t)$, and multiplying by the cross-sectional area of the diode to obtain the current in amperes:

$$I_p = (q A D_p p_{n0} / W_n)[\exp(V_a/V_t) - 1] \tag{3.7}$$

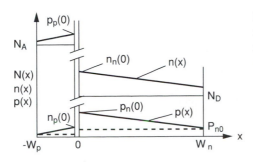

Figure 3.3 Free-carrier distributions in the P^+N narrow-base diode. The depletion layer is assumed narrow compared to the neutral region widths W_n, W_p.

By analogy, the minority carrier electron current flowing in the *P* side is given by:

$$I_n = (qAD_n n_{p0}/W_p)[\exp(V_a/V_t) - 1] \tag{3.8}$$

Note that the electrons in the *P–N* diode of Fig. 3.3 flow to the left, giving a conventional current flowing from left to right. The holes, being positive charge carriers, give a conventional current, which also flows from left to right. The two currents thus add together. In order to see the relative importance of the electron and hole current, let us examine the ratio:

$$I_p/I_n = (D_p p_{n0} W_p)/(D_n n_{p0} W_n) \tag{3.9}$$

Using the fact that $p_{n0} = n_i^2/N_D$ and $n_{p0} = n_i^2/N_A$, gives:

$$I_p/I_n = (D_p N_A W_p)/(D_n N_D W_n) \tag{3.10}$$

For a *P⁺N* diode the *P*-region doping is of order 10^{19} cm^{-3} and the *N*-region doping of order 10^{15} cm^{-3}. Even if the region widths differ by a factor of 10, it is clear that the hole current in a *P⁺N* diode is very much larger than the electron current. We conclude that in all practical diodes, made by diffusing a heavily doped layer into a moderately doped background layer, *it is the lightly doped side that determines the current–voltage characteristics.* Thus in an *N⁺P* diode the current is predominantly due to electrons injected into the *P* region. In general we can write:

$$I = I_0[\exp(V_a/V_t) - 1] \tag{3.11}$$

where I_0 (frequently also referred to as I_s and called the reverse leakage current or reverse saturation current) is given by the above expression for either hole or electron current.

$$P^+N : \quad I_0 = qAD_p p_{n0}/W_n \tag{3.12a}$$
$$N^+P : \quad I_0 = qAD_n n_{p0}/W_p \tag{3.12b}$$

It should be noted that the reverse saturation current I_0 does not depend on the potential barrier (contrary to the forward *I–V* characteristic). I_0 is due to the small diffusion gradient that exists in reverse bias. This is examined in more detail in Section 3.7, where we will see that other more important components contribute to the actual value of I_0 in a real diode.

Figure 3.4 shows the *I–V* characteristic corresponding to the above result. The current is very small indeed under reverse bias and increases rapidly when the forward bias exceeds typically 0.5 V.

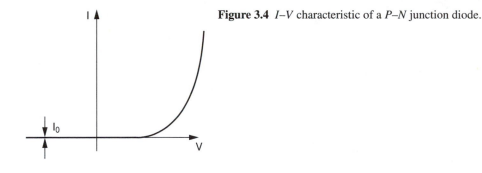

Figure 3.4 *I–V* characteristic of a *P–N* junction diode.

For example, consider a P^+N diode of area 1 cm^2 (typical of a high-current rectifier diode), with the lightly doped region 50 μm thick and doped 10^{14} cm^{-3}. We find the value of I_0 from Eq. (3.12a). Assume $D_p = 10$ cm^2/s [from Fig. 2.11 and Eq. (2.31)]; $n_i^2 = 2 \times 10^{20}$.

$I_0 = 1.6 \times 10^{-19} \times 1 \times 10 \times (2 \times 10^{20}/10^{14})/50 \times 10^{-4} = 6.4 \times 10^{-10}$ or 0.64 nA (1 nA = 10^{-9} A). At a forward bias of 0.6 V, the current is approximately 7 A (using $V_t = 0.0259$ V).

Note that we only give approximate values, since both I_0 and the forward current are very sensitive to the operating junction temperature because of the presence of n_i^2, which, from the basic result given in Chapter 2, is proportional to $\exp(-E_g/kT)$. Taking T varying from 27 to 17°C and using the band gap of silicon (1.1 eV) gives a reduction in I_0 of approximately 1/5 for this 10°C change in temperature.

3.3 THE HIGH-CURRENT *PIN* DIODE

This is a much more practical diode and is used for nearly all rectifier applications, particularly when a high reverse operating voltage is required. In its high-current form, an entire wafer is often used. This is made with a very light doping level, ideally undoped, the thickness being of order 100 m. A shallow (several microns deep) *P* diffusion is made on the top and an *N* diffusion on the bottom of the wafer. Figure 3.5 shows the doping distribution for an idealized case with constant impurity concentration at both ends and a near-intrinsic (light *N*-type doping) region extending over the thickness W_i. Note that it is very difficult to obtain residual doping levels less than about 10^{13} cm^{-3}. The center layer can be either lightly *N* or *P* type, without affecting the following discussion, which is restricted to the case of strong forward bias. Let us discuss the ideal case with zero center-layer doping. The value of injected electron concentration on the intrinsic side of the *N–I* junction can be found from the injection law, with n_i replacing n_{p0}. Likewise the injected hole concentration on the intrinsic side of the *P–I* junction can be found. The results are:

$$p(0) = n_i \exp(V_{a1}/V_t) \tag{3.13}$$

$$n(W_i) = n_i \exp(V_{a2}/V_t) \tag{3.14}$$

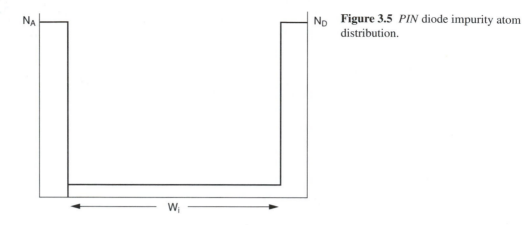

Figure 3.5 *PIN* diode impurity atom distribution.

where V_{a1} and V_{a2} are the parts of the total applied bias appearing across each junction. Since the structure is symmetrical, the applied bias V_a will split equally across each junction, giving:

$$V_{a1} = V_{a2} = V_a/2 \tag{3.15}$$

Figure 3.6 illustrates this situation. The concentration of electrons and holes is the same at every point within the center layer over the distance W_i. This must be the case, since we are dealing with a neutral region (any space-charge layer is confined to a very narrow region near each forward-biased junction), and the total negative charge concentration due to the electrons must be balanced by the positive charge due to the holes. Since the forward bias across each junction is the same, it follows that $p(0) = n(W_i)$ (note that in a real *PIN* diode, the distribution is somewhat skewed due to both the finite center-layer doping giving rise to an unequal bias across each junction, and also to unequal electron and hole mobilities). (See also CD-ROM BIPGRAPH example DIOD2PIN; "Minority carrier conc. vs depth.")

The current is due entirely to the free-carrier charge within the W_i region recombining with a lifetime τ_i. Using the result for recombination current and the charge continuity equation in Chapter 2, Eq. (2.87), we can write:

$$I_{rec} = Q/\tau_i \tag{3.16}$$

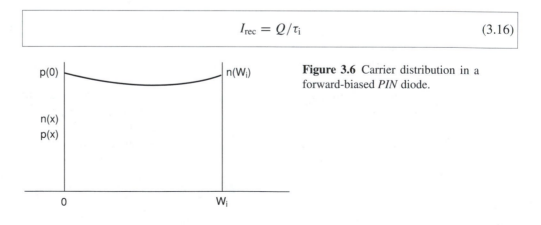

Figure 3.6 Carrier distribution in a forward-biased *PIN* diode.

(The hole current flowing into the N^+ region is small, tending to zero if the N^+ region doping is high; conversely for the electron current flowing into the P^+ region, hence the direct applicability of the charge continuity equation.) Recombination tends to make the hole and electron concentrations decay moving away from the points of injection. However, the value of the total charge Q may be readily approximated if we consider the case where the center-layer thickness does not exceed a carrier diffusion length L_i. In fact, it turns out that $W_i \approx L_i$ is a near-optimal design condition for many applications. Under the condition $W_i \leq L_i$, Q may safely be approximated by a rectangular distribution and hence we have:

$$I_{rec} = q A p(0) W_i / \tau_i = q A n(W_i) W_i / \tau_i \qquad (3.17)$$

Substituting for the bias dependence of $p(0)$ gives the I–V characteristic:

$$I = I_{0pin} \, \exp(V_a / 2V_t) \qquad (3.18)$$

where

$$I_{0pin} = q A n_i W_i / \tau_i \qquad (3.19)$$

This result is closely observed in practice, as long as the forward bias is sufficiently high such that the injected carrier concentrations considerably exceed the (low) background doping level.

3.4 HIGH-LEVEL INJECTION

Rather remarkable is the fact that any diode operating under high-current conditions exhibits the same $\exp(V_a / 2V_t)$ law. The change in slope on a $\log(I)$ versus linear V_a plot occurs when the injected carrier concentration exceeds the background doping level. This condition is known as *high-level injection*, or HLI for short.

Figure 3.7 shows the narrow-base diode carrier distribution under high-level injection. The applied bias voltage V_{aHLI} corresponding to the onset of high-level injection may be obtained approximately from the injection law, by setting $p(0)_{HLI} = N_D$, giving:

$$V_{aHLI} = V_t \, \ln[p(0)_{HLI}/p_{n0}] = V_t \, \ln(N_D^2/n_i^2) \qquad (3.20)$$

or

$$V_{aHLI} = 2V_t \, \ln(N_D/n_i) \qquad (3.21)$$

For a doping level $N_D = 10^{16}$ cm^{-3}, at room temperature, V_{aHLI} is approximately 0.8 V. The I–V characteristic may be derived by taking into account the electric field in the neutral region and the corresponding potential drop (which must be supplied by part of the applied voltage V_a). We include this analysis as an exercise for the interested reader (Problem 5).

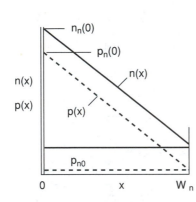

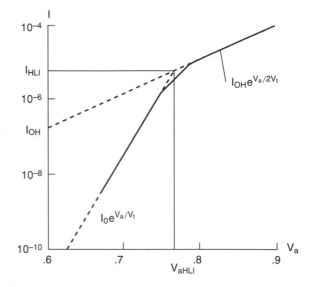

Figure 3.7 Carrier distributions in a *P⁺N* narrow-base diode under high-level injection.

Figure 3.8 Current–voltage characteristic of a narrow-base or *PIN* diode on a log–linear plot at moderate and high forward bias. This illustrates the change from low-level to high-level injection.

The overall current–voltage characteristic of a narrow-base diode or a *PIN* diode is as shown in Fig. 3.8 on a log–linear plot, where the change of slope is readily identified. The current–voltage equation for a diode is often written as:

$$I = I_0[\exp(V_a/mV_t) - 1] \tag{3.22}$$

where the factor m assumes a value between 1 and 2 depending on the bias range under observation. It is important to note that the value of I_0 must also be chosen carefully and depends on the bias range for which this simple "model" is to be used. This is obvious from the log–linear plot of Fig. 3.8, where the value of I_{OH} (I_{0pin} for the *PIN* diode) from the $m = 2$ region is in this case four orders of magnitude greater than the I_0 value for low-level injection. It is also possible from a careful study of this plot to determine the current I_{HLI} at which the slope changes (gradually) from $m = 1$ to 2. We shall see later in this chapter that there exists a second situation, at very low forward bias, where m can again approach a value equal to 2.

3.5 OTHER DIODES

3.5.1 The wide-base diode

Figure 3.9 shows the carrier distributions in a *P⁺N* diode in which each side has a thickness (or width) much greater than a diffusion length, that is, $W_n \gg L_p$, where $L_p = \sqrt{(D_p\tau_p)}$ on

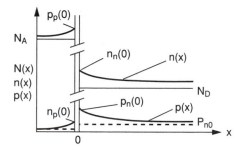

Figure 3.9 Carrier distributions in a P^+N wide base diode.

the N side. $p_n'(x)$ is used to denote the excess hole concentration $p_n(x) - p_{n0}$. In Appendix I we show that the distribution of holes versus distance on the N side is an exponential decay with characteristic length L_p given by:

$$p_n'(x) = p_n'(0) \exp(-x/L_p) \tag{3.23a}$$

The hole current at $x = 0$ on the N side is therefore:

$$I_p = -qAD_p \left.\frac{dp}{dx}\right|_{x=0} = qAD_p p_n'(0)/L_p \tag{3.23b}$$

The final current–voltage relationship is obtained as with the narrow-base diode [Eqs. (3.6)–(3.7)] by using the injection law to obtain:

$$I = I_0[\exp(V_a/V_t) - 1] \tag{3.24a}$$

where for the P^+N structure shown, I_0 is made up mainly of holes injected into the moderately doped N region and is therefore given to a good approximation by:

$$I_0 = qAD_p p_{n0}/L_p \tag{3.24b}$$

where $p_{n0} = n_i^2/N_D$.

The wide-base diode is of little practical importance, since at high currents the voltage drop in the (by definition) wide neutral regions makes the exponential I–V characteristic tend to a linear resistance, with poor rectification performance.

3.5.2 The P^+NN^+ diode

This structure is really the *PIN* diode operating under low-level injection, since as already mentioned, zero center-layer doping levels are practically unobtainable. Since normal operation puts the diode into high-level injection, where the *PIN* theory applies, we shall not devote space here to a complete derivation of the I–V characteristic. This is given in Appendix J for completeness.

Figure 3.10 shows the impurity profile (a more realistic drawing of the idealized impurity profile of Fig. 3.5) and hole concentration distribution.

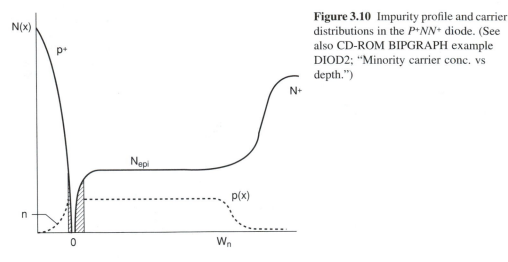

Figure 3.10 Impurity profile and carrier distributions in the *P+NN+* diode. (See also CD-ROM BIPGRAPH example DIOD2; "Minority carrier conc. vs depth.")

Notice that the presence of the *NN+* "junction" forces the carriers down towards zero at $x = W_n$ (in simple terms, this is because on the right-hand side of the *N–N+* junction, the hole concentration is n_i^2/N_D^+, and for N_D^+ very high, typically of order 10^{19} cm^{-3}, this yields a hole concentration much less than that in the center *N* layer). The charge distribution is thus nearly rectangular but with a slight dip due to the finite amount of recombination. The current, for the case where the center-layer region thickness W_n is not greater than a hole diffusion length, is:

$$I = I_0[\exp(V_a/V_t) - 1] \tag{3.25}$$

where I_0 is made up primarily of holes injected into the *N* region, recombining with a lifetime τ_p and is given by:

$$I_0 = q A p_{n0} W_n/\tau_p \tag{3.26}$$

3.5.3 Narrow-base diode with recombination

Quite often a diode will not fit any of the above categories. A not uncommon case is a diode that is neither a wide-base nor strictly a narrow-base diode. Consider the situation shown in Fig. 3.11, where we have shown the *N* region of a *P+N* "narrow-base diode," in which the *N*-region thickness W_n is not so small that recombination can be neglected; that is, W_n is not small enough compared to the hole diffusion length L_p. The hole concentration distribution will be as shown by the broken curve (b). Recombination forces the hole distribution to decay more rapidly from the point of injection ($x = 0$). The magnitude of the hole concentration gradient is increased at $x = 0$, and hence the current is increased. A good estimate of the

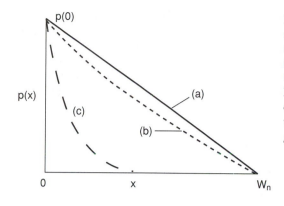

Figure 3.11 P^+–N narrow-base diode with recombination: hole distribution in the N region: (a) no recombination; (b) finite recombination; (c) recombination with $L_p \ll W_n$. (See also CD-ROM BIPGRAPH examples DIOD1, DIOD5, DIODP1, DIODP03; "Minority carrier conc. vs depth." This generates plots for diodes with base region carrier lifetimes of 1 μs, 0.5 μs, 0.1 μs, 0.05 μs.

current due to recombination can be made by assuming the distribution to be triangular, and taking the corresponding charge Q_p. This gives:

$$Q_p = \tfrac{1}{2} q A p_n(0) W_n \tag{3.27}$$

The current due to recombination of this charge with a lifetime τ_p is thus:

$$I_{\text{rec}} = Q_p / \tau_p \tag{3.28}$$

This may be compared to the original result for narrow-base diode hole current [Eqs. (3.6) and (3.7)] by taking the ratio:

$$I_{\text{NBD}}/I_{\text{rec}} = [q A D_p p_n(0)/W_n] / \left[\tfrac{1}{2} q A p_n(0) W_n / \tau_p\right] \tag{3.29}$$

$$= 2 D_p \tau_p / W_n^2 \tag{3.30}$$

Substituting for hole diffusion length $L_p = \sqrt{(D_p \tau_p)}$ gives:

$$I_{\text{NBD}}/I_{\text{rec}} = 2(L_p/W_n)^2 \tag{3.31a}$$

and the total current is:

$$I_{\text{NBD total}} = I_{\text{NBD}} \left[1 + \tfrac{1}{2}(W_n/L_p)^2\right] \tag{3.31b}$$

This is strictly valid only for $W_n \ll L_p$ but gives reasonable results up to $W_n = L_p$. If the hole diffusion length is three times the N region width, the recombination current is thus about 5% of the theoretical narrow-base diode current. In the extreme case where the lifetime is so small that the diffusion length $L_p \ll W_n$, the diode behaves as a wide-base diode, as shown by curve (c) of Fig. 3.11.

3.6 LOW FORWARD BIAS

Ideally, in all the above diodes, the reverse current is constant at the value given by the I_0 value. However, in practice, all silicon diodes operating at room temperature exhibit an I–V characteristic dominated at low forward bias by what is referred to as *space-charge recombination current*. Figure 3.12 shows the carrier distribution in the space-charge layer of a

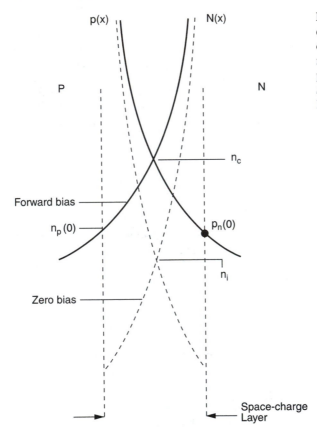

Figure 3.12 Hole and electron distributions within the space-charge layer of a *P–N* diode under moderate forward bias. Reprinted with permission from D. J. Roulston *Bipolar Semiconductor Devices*, McGraw–Hill, New York, 1990.

symmetrical *P–N* junction (chosen here for its pictorial simplicity and ease of understanding). The injected carrier concentrations at each side of the space-charge layer have already been studied. Let us focus on the hole distribution.

$$p_n(0) = p_{n0} \exp(V_a/V_t) \tag{3.32}$$

The hole concentration on the *P* side is, for low forward bias, the same as the acceptor concentration; that is, $p_p(0) = N_A$. The quantities can better be visualized on a log–linear plot, as shown in Fig. 3.13. The shaded regions represent excess minority carrier charge.

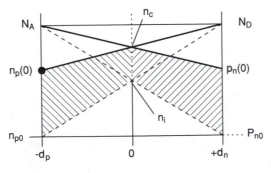

Figure 3.13 Carrier distributions in the space-charge layer using a logarithmic vertical scale.

The current due to this charge with an equivalent carrier lifetime τ is:

$$I_{\text{scr}} = Q_p/\tau_p \tag{3.33}$$

where Q_p is proportional to the free-carrier concentration n_c at the junction. Let us examine how n_c varies with applied bias. On the N side of the space-charge layer $p(0) = p_{n0} \exp(V_a/V_t)$. On the P side $p(0)$ is constant. A good estimate of n_c at the junction is given by taking the geometric mean:

$$n_c = [p(0)N_A]^{1/2} \tag{3.34}$$

Substituting for $p(0)$ and setting $N_D = N_A$, for the symmetrical junction under consideration gives:

$$n_c = n_i \exp(V_a/2V_t) \tag{3.35}$$

This result could also have been obtained quite rigorously by applying the Boltzmann relations inside the space-charge layer. The current may thus be expressed in the form:

$$I_{\text{scr}} = I_{0r} \exp(V_a/2V_t) \tag{3.36}$$

In Appendix K, it is shown that:

$$I_{0r} = 2q\, A n_i x_c/\tau \tag{3.37}$$

where $x_c = d_n V_t/(V_{bi} - V_a)$ is the characteristic length of the exponential decay of carriers inside the space-charge layer. An exact analysis using an idealized recombination model indicates that $\tau = \sqrt{(\tau_p \tau_n)}$.

We see that the result is a space-charge recombination current that varies as $\exp(V_a/2V_t)$. While we have made some simplifying assumptions in the above analysis, including the use of an effective lifetime τ, the result is remarkably close to the experimental characteristic observed on a wide range of silicon diodes. In practice, depending on the nature of the recombination process in the space-charge layer, this space-charge recombination current is found to vary as $\exp(V_a/mV_t)$, where m is typically between 1 and 2.

The complete forward-bias I–V characteristic of a silicon diode is thus as shown in Fig. 3.14. The three regions with m values of 2, 1, 2 may not always be identifiable. Depending on the relative magnitudes of the different current components, an "average" m value of between 1 and 2 may often be used as a best fit. At very high currents all diodes exhibit a departure from the exponential characteristic; this is due to series resistance made up of contact resistance and resistance in various regions of the semiconductor device.

3.7 REVERSE BIAS

3.7.1 Moderate reverse bias

Under reverse-bias conditions the law of injection for holes on the N side gives:

$$p_n(0) = p_{n0} \exp(V_a/V_t) \tag{3.38}$$

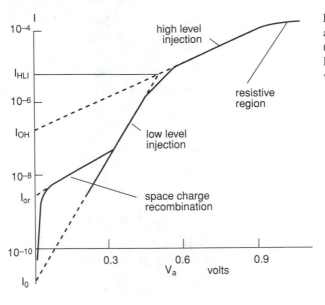

Figure 3.14 *I–V* characteristic of a *PN* diode on a log–linear scale. (See CD-ROM BIPGRAPH example DIODIV; "J-Va and m-Va" [in DOS version type DIV DIODIV].)

For V_a more than a few tenths of a volt negative, the injected hole concentration is for all practical purposes zero. A similar result holds for electrons on the *P* side. The free-carrier concentration within the space-charge layer is thus depressed down to a value almost zero. However, we now have a mechanism that is the opposite of recombination; it is called *generation* and is governed by the same time constants, the electron and hole lifetimes. Since the space-charge layer is depleted of carriers, and since the donor and acceptor atoms have "lost" their corresponding electrons and holes, this region "looks like" a pure intrinsic region in which the carrier concentration would be n_i, if not forced to a value of zero by the reverse bias. This is shown in Fig. 3.15. The net result is that generation tries to bring the carrier concentration up to the n_i value, and produces a corresponding *space-charge generation current*:

$$I_{scrgen} = -Q/\tau \tag{3.39a}$$

where Q is the charge that would exist due to n_i. This charge is is simply $qAn_i(d_p + d_n)$; hence:

$$I_{scrgen} = qAn_i(d_p + d_n)/\tau \tag{3.39b}$$

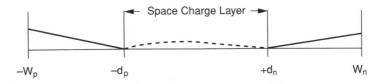

Figure 3.15 Carrier distributions in a reverse-biased *PN* narrow-base diode. The space-charge layer extends from $-d_p$ to $+d_n$.

Note that d_p and d_n are each proportional to the square root of the total junction voltage $(V_{bi} - V_a)$ and therefore increase with increasing reverse bias as in Eq. (2.54) or (2.62). A rigorous analysis indicates that the effective lifetime in this scenario is the sum of the electron and hole lifetimes: $\tau = \tau_p + \tau_n$. Figure 3.15 is drawn for a *PN* narrow-base diode, but this space-charge generation current will add to the neutral-region currents, as determined for each particular diode.

3.7.2 Large reverse bias: avalanche multiplication and breakdown

As the reverse bias is increased, a new phenomenon occurs called *avalanche multiplication*. For a depletion layer thickness of 1 μm with 10 V across it, the peak electric field is 2×10^5 V/cm. At this high value of field, the carriers in the space-charge layer are accelerated to a sufficiently high velocity that when a collision occurs with a lattice atom, an electron can be liberated, thereby creating an electron–hole pair. The new electron and hole drift in opposite directions, and are again accelerated. The scenario is illustrated in Fig. 3.16. If the field is high enough and if the distance is large enough, the degree of multiplication increases indefinitely, and a large current flows. The maximum reverse-bias voltage that can be applied before the current tends to infinity is called the *avalanche breakdown voltage* V_{br}. The current increases gradually by an avalanche multiplication factor M for reverse voltages V less than V_{br} according to the empirical law:

$$M = 1/[1 - V/V_{br})^n]$$ (3.40)

where n for silicon diodes lies between 1 and 3.

Calculation of reverse breakdown voltage

Breakdown occurs at a value of peak electric field E_{br}, which is roughly constant at 3×10^5 V/cm for moderate doping levels (the variation of E_{br} with doping level will be discussed later). The breakdown voltage of the P^+N junction can be obtained from the solution to Poisson's equation for total N-side voltage and peak electric field obtained in Eq. (2.52).

$$V_{jtot} = (q/2\epsilon)N_D d_n^2$$

Substituting for depletion layer thickness d_n in terms of peak electric field from Eq. (2.48), setting the field equal to E_{br}, and calling the resulting breakdown voltage V_{br} gives:

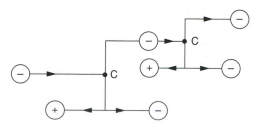

Figure 3.16 Multiplication of carriers due to impact ionization at high reverse bias.

$$V_{br} = \epsilon E_{br}^2 / 2q N_D \tag{3.41}$$

Example of breakdown voltage calculation: For a doping level on the lightly doped side of 10^{16} cm^{-3}

$$V_{br} = \frac{10^{-12} \times 9 \times 10^{10}}{2 \times 1.6 \times 10^{-19} \times 10^{16}} = 28 \text{ V}$$

The corresponding thickness of the N_D layer from Eq. (2.62) is:

$$W_n = d_n = \sqrt{\frac{2\epsilon \times 28}{q \times 10^{16}}} = 1.9 \, \mu\text{m}$$

The above example shows that we can choose the thickness and doping level of the lightly doped side of a diode for a given peak inverse voltage. A larger thickness would not be used because of the additional series resistance, which would adversely affect the forward-bias characteristics at high current. Note that the above values of voltage include the built-in barrier potential.

For a lightly doped *PIN* diode as in Fig. 3.5 the field is almost constant with respect to distance under high reverse bias. In this case the breakdown voltage is given by: $V_{br} = E_{br} W_i$ and is independent of doping level.

For high doping levels on both sides of the junction, the space-charge layer becomes very narrow, and the field reaches a value of 10^6 V/cm at low reverse, or even at zero bias. In this situation, the distance traveled by the carriers is not sufficient to create the above impact ionization process. A different mechanism comes into play, called *Zener breakdown*. This may be thought of as direct disruption of the covalent bonds, thereby creating free electrons. The avalanche mechanism dominates for breakdown voltages in excess of 6 V. Below this value, the Zener mechanism takes over. An interesting situation occurs for doping levels in excess of about 10^{19} cm^{-3} on both sides of the junction. Let us calculate the built-in barrier potential and the peak electric field for this case.

Example of depletion layer properties for high doping (10^{19} cm^{-3}) on both sides: Equation (2.43) gives the built-in barrier voltage:

$$V_{bi} = V_t \, \ln(N_A N_D / n_i^2) = 0.025 \, \ln(10^{19} \times 10^{19} / 10^{20}) \approx 1.0 \, \text{V}.$$

The depletion layer thickness on one side (for one-half the barrier voltage V_{bi}) is from Eq. (2.54):

$$d_n = [\epsilon V_{bi}/(q N_D)]^{1/2} = 0.008 \, \mu m$$

The peak electric field is

$$E_{max} = (q/\epsilon) N_D d_n = 1.3 \times 10^6 \text{ V/cm}$$

This example shows that for high doping on both sides of the junction, the junction is in Zener breakdown even at zero bias, due to the electric field produced by the built-in barrier potential. As the doping level is increased further, to, say, 10^{20} cm^{-3}, the diode will only have a field lower than 10^6 V/cm if a sufficient forward bias is applied. This is the tunnel diode, which exhibits a negative differential resistance and may thus be used for amplifier applications. The various I–V characteristics corresponding to a progressive increase in doping level on both sides of the junction are illustrated in Fig. 3.17.

The complete current–voltage diode characteristic

Figure 3.18 shows the complete current–voltage characteristic of a diode using linear scales. For most applications the significant parameters are: (1) peak inverse breakdown voltage; (2) forward voltage drop at maximum rated current (usually just before the onset of the high-current resistive region at just under 1.0 V for a silicon diode), (3) the value of this maximum current. It may be noted that the breakdown voltage depends only on the doping level and thickness of the lightly doped side of the diode; the maximum current (for a given forward voltage drop) depends on the doping level of the lightly doped side and on the cross-sectional area.

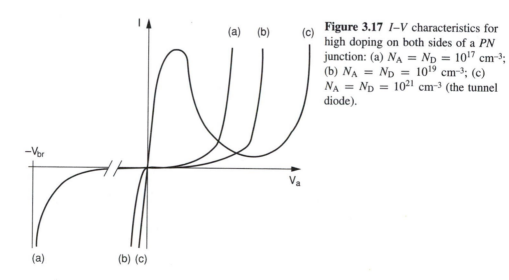

Figure 3.17 I–V characteristics for high doping on both sides of a PN junction: (a) $N_A = N_D = 10^{17}$ cm^{-3}; (b) $N_A = N_D = 10^{19}$ cm^{-3}; (c) $N_A = N_D = 10^{21}$ cm^{-3} (the tunnel diode).

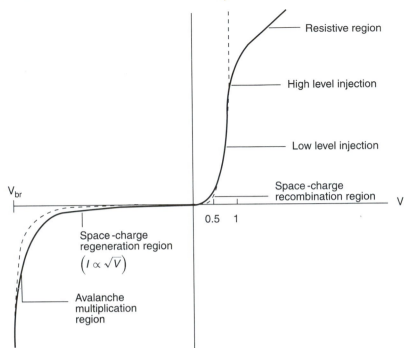

Figure 3.18 Complete diode *I–V* characteristics using linear scales. Reprinted with permission from D. J. Roulston, *Bipolar Semiconductor Devices*, McGraw–Hill, New York, 1990.

3.8 DIODE SMALL-SIGNAL CAPACITANCE AND EQUIVALENT CIRCUITS

3.8.1 Depletion-layer capacitance

The space-charge layer of a diode, reverse or forward biased, contains a charge Q on either side, of opposite sign. The value of this charge is given by:

$$Q = qAN_A d_p = qAN_D d_n \tag{3.42}$$

If the bias voltage is increased by an amount dV, the charge increases on each side by an amount dQ. This is illustrated in Fig. 3.19 for a P^+N diode. The situation is clearly very similar to that pertaining in a parallel-plate capacitor in which the incremental charge gets added on each plate. In this case the conventional formula for capacitance is $C = Q/V = \epsilon A/d$, where d is the distance separating the two parallel plates. However, in the $P–N$ junction, the distance separating the edges of the depletion layer is a function of applied voltage. We can use the parallel-plate capacitance formula to write the depletion layer capacitance of the junction:

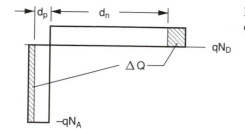

Figure 3.19 Incremental charge increase in a *P+N* diode for depletion layer capacitance calculation.

$$C_j = \epsilon A / (d_p + d_n) \tag{3.43}$$

However, it is important to observe that this is a *small-signal nonlinear capacitance* and is a function of applied voltage. It is defined by:

$$C_j = \frac{dQ}{dV} \tag{3.44}$$

The interested reader can show that this leads to exactly the above "parallel-plate capacitance" result. This capacitance can be measured by applying a dc bias to the diode in series with a small sinusoidal voltage. By measuring the alternating current, the capacitance is obtained. To make the measurement accurately, a radio frequency signal is used (at least 100 kHz in order to measure the capacitive current).

For the *P+N* diode $d_p \ll d_n$ and hence the capacitance becomes:

$$C_j = \epsilon A / d_n \tag{3.45}$$

We have already determined d_n as a function of total junction voltage $V_{jtot} = V_{bi} - V_a$ [Eq. (2.62)]:

$$d_n = [2\epsilon V_{jtot}) / (q N_D)]^{1/2} \tag{3.46}$$

The capacitance may thus be written in the form:

$$C_j = \epsilon A / [2\epsilon V_{jtot}) / (q N_D)]^{1/2} \tag{3.47}$$

or, more conveniently:

$$C_j = C_0 / (1 - V_a / V_{bi})^{1/2} \tag{3.48}$$

where C_0 is the value of C_j for $V_a = 0$, obtained by setting $V_{jtot} = V_{bi}$ in the above equations.

This inverse square root dependence of depletion layer (or junction) capacitance on total junction voltage is a well-established law for any junction with constant doping on the lightly doped side. Figure 3.20 shows the $C(V)$ law on both a linear and a log–log plot.

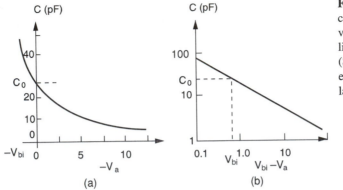

Figure 3.20 Depletion layer capacitance of an abrupt junction versus applied bias voltage: (a) linear axes; (b) logarithmic axes. (See also CD-ROM BIPGRAPH example DIOD30; "Depletion layer capacitance vs Vjtotal.")

Note that the exponent -0.5 is the slope of the log–log graph. Note also that in order to draw the log–log plot, the value of the built-in barrier voltage V_{bi} must be known. In practice a value of 0.7 V may be used for most silicon diodes with little loss of accuracy. Alternatively, by plotting $1/C^2$ versus V_a, the intercept on the x axis gives V_{bi}.

For a linear graded junction, where from Eq. (2.67) the total voltage is proportional to d_n^3, the capacitance varies proportionally to $V_{jtot}^{-1/3}$. A capacitance versus voltage plot can be used to determine important information about the nature of the impurity profile $N(x)$.

The voltage dependence of the junction capacitance has some direct applications, in the varactor diode, used for tuning purposes (a dc voltage applied to the diode can be used to alter the resonant frequency in an oscillator circuit), or for frequency multiplication.

3.8.2 Diffusion capacitance

There is a second capacitance associated with a P–N junction; this is due to the minority carrier charge in each neutral region. Figure 3.21 shows the hole distributions in the N region of a P^+N-junction narrow-base diode for a bias V_a and a bias $V_a + dV_a$. If the P side is much more heavily doped than the N side, we may safely neglect the electron current and charge on the P side. The excess charge Q_p due to the triangular distribution of holes is:

$$Q_p = \tfrac{1}{2} q A [p_n(0) - p_{n0}] W_n \tag{3.49}$$

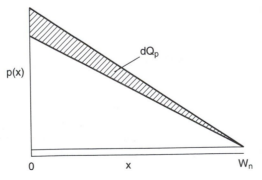

Figure 3.21 Hole charge (for applied voltages V_a and $V_a + dV_a$) on the N side of a P^+N narrow-base diode for diffusion capacitance calculation.

where $p_n(0) = p_{n0} \exp(V_a/V_t)$.

We can define the *small-signal diffusion capacitance* due to this charge as:

$$C_{\text{diff}} = \frac{dQ_p}{dV_a} \tag{3.50}$$

This may conveniently be written as:

$$C_{\text{diff}} = \frac{dQ_p}{dI} \frac{dI}{dV_a} \tag{3.51}$$

Since the current I and the charge Q_p are each proportional to $\exp(V_a/V_t)$, the derivative dQ_p/dI is equal to Q_p/I. This ratio has the dimensions of charge/current = time. The time constant t_b is given by:

$$t_b = Q_p/I = \tfrac{1}{2}qA[p_n(0) - p_{n0}]W_n/\{qAD_p[p_n(0) - p_{n0}]/W_n\} \tag{3.52}$$

or

$$t_b = W_n^2/2D_p \tag{3.53}$$

This may be shown to be the time taken for the holes to diffuse across the distance W_n and is called the *base transit time* of the diode.

The term dI/dV_a is the slope of the exponential I–V characteristic at a given current I as shown in Fig. 3.22. This is the small-signal conductance g_s of the diode at a current I. One can also define $r_d = 1/g_s$ as the small-signal resistance. We include m in the dc characteristic ($1 < m < 2$) and use I_0' where I_0' can be either I_0 or I_{OH} in Fig. 3.8 or I_{0r}, Eq. (3.37), to make the result for g_s more general.

$$I = I_0'[\exp(V_a/mV_t) - 1] \tag{3.54}$$

$$g_s = \frac{dI}{dV_a} = (I_0'/mV_t)\,\exp(V_a/mV_t) \tag{3.55}$$

This may be simplified for normal forward bias [V_a greater than 0.1 V such that $\exp(V_a/mVt) \gg 1$] by substituting for the current I, to give:

$$g_s = 1/r_d = I/mV_t \tag{3.56}$$

The diffusion capacitance may thus be written as:

$$C_{\text{diff}} = g_s t_b \tag{3.57}$$

This analysis has neglected the fact that not all charge is "recovered" when an ac signal is applied. Some carriers diffuse to the ohmic contact at W_n and are "lost". An exact analysis shows that a reduction of a factor $\tfrac{2}{3}$ applies to the value of C_{diff} given by Eq. (3.57).

For a *PIN* diode one may pursue a similar analysis. Both the free-carrier charge and the current are proportional to the injected carrier concentration. It is a simple matter to show that

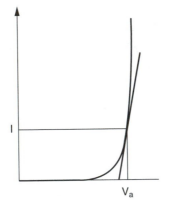

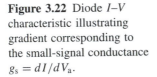

Figure 3.22 Diode *I–V* characteristic illustrating gradient corresponding to the small-signal conductance $g_s = dI/dV_a$.

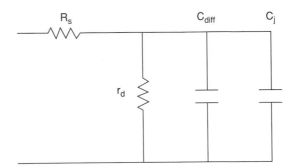

Figure 3.23 Small-signal equivalent circuit of the diode.

the charge-to-current ratio Q/I in this case is equal to the carrier lifetime τ. The diffusion capacitance is then given by:

$$C_{\text{diff}(PIN)} = g_s\tau \tag{3.58}$$

Similar results are also obtained for the wide-base and P^+NN^+ diodes.

Diode small-signal equivalent circuit

Based on the above results, we can draw a small-signal equivalent circuit for a diode biased at some voltage V_a, corresponding to a current I. This is shown in Fig. 3.23 and is a convenient way to calculate the response of the diode to a small sinusoidal current or voltage at a given frequency. We have added a resistance R_s to account for series resistance due to the bulk semiconductor material and the resistance of the metal–semiconductor contacts. Note that under reverse bias, only R_s and C_j remain in the equivalent circuit since r_d becomes infinite ($g_s = 0$) and C_{diff} becomes zero.

Since the diffusion capacitance is proportional to the dc current, it varies as $\exp(V_a/V_t)$, this will always dominate over the depletion-layer capacitance at large forward bias. Below about 0.5 V the depletion-layer capacitance C_j becomes greater than C_{diff}.

3.9 DIODE SWITCHING BEHAVIOR

Let us consider first the case of a diode switched by abruptly applying a forward voltage step, as shown in Fig. 3.24(a). The corresponding voltage and current waveforms are shown in

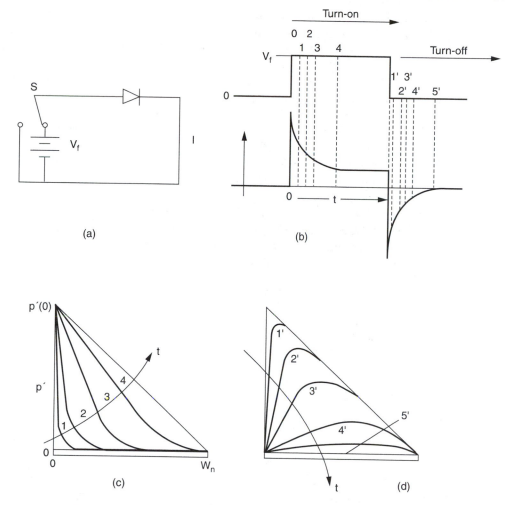

Figure 3.24 Diode switching from a voltage source: (a) circuit; (b) voltage and current waveforms; (c) hole distribution during turn-on; (d) hole distribution during turn-off (for narrow-base diode). After Roulston [7]. Reprinted with permission from *Bipolar Semiconductor Devices*, McGraw–Hill, New York, 1990.

Fig. 3.24(b). The remaining two diagrams show the hole concentration distributions at various times in the N side of a P^+–N narrow-base diode.

In this ideal case with zero series resistance, the hole concentration $p_n(0)$ is forced at time $t = 0$ to the value given by the injection law: $p_n(0) = p_{n0} \exp(V_a/V_t)$. Since it takes a finite time for the holes to diffuse across the N region, the hole concentration gradient will initially be a very large negative value (infinite if there is no series resistance), as shown in (c) curve 1. The current will thus start with a large positive spike as shown in (b) (recall that $I_p = -qAD_p\, dp/dx$). As the holes diffuse across the N region, the concentration gradient gradually decreases towards the steady-state value (the triangular distribution already discussed), and the current decays to its dc value given by the diode *I–V* law. The time taken to establish this steady-state condition will be comparable to the diffusion transit time t_b across the *N* region, where $t_b = W_n^2/2D_p$. The curves in Fig. 3.24(c) are labeled 1, 2, 3, 4 corresponding to the times 1, 2, 3, 4 in Fig. 3.24(b).

If the voltage is now brought abruptly to zero, the reverse situation occurs. The hole concentration $p_n(0)$ falls abruptly to zero, thereby creating a large positive gradient and a large negative current spike. The hole concentration and the current then gradually decay towards their equilibrium values. The labels $1'$, $2'$, $3'$, $4'$, $5'$ in Fig. 3.24(d) correspond to the times marked on Fig. 3.24(b).

In practice a diode circuit will have some series resistance, and it is more useful to determine the switching characteristics for the case where the current is limited by this resistance. Fig. 3.25 shows the circuit, waveforms, and hole distributions for this case. Providing the source voltage is at least a few volts, substantially greater than the diode "on" voltage of about 0.7 V, the current during turnon will be approximately V_a/R. The hole concentration gradient will now be approximately constant (since $I_p = -qAD_p\,dp/dx$). The turnon time will again be comparable to the diffusion transit time t_b across the *N* region.

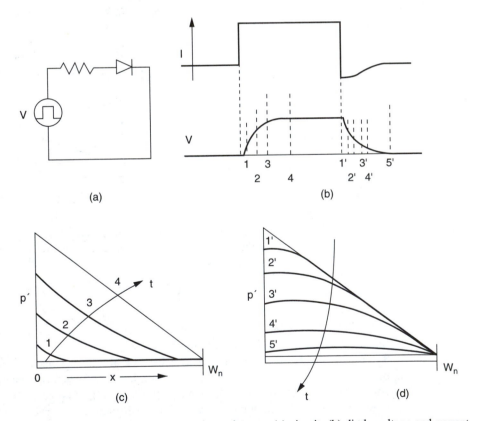

Figure 3.25 Diode switching through a series resistance: (a) circuit; (b) diode voltage and current waveforms; (c) hole distributions during turn-on; (d) hole distributions during turn-off. After Roulston [7]. Reprinted with permission from *Bipolar Semiconductor Devices*, McGraw–Hill, 1990.

Note that when the source voltage is brought to zero (after the diode has reached steady-state forward bias), the current actually reverses. This is because the voltage across the diode only decays relatively slowly from its value of about 0.7 V towards zero; so the reverse current starts at a value of about $-0.7/R$ and decays towards zero.

We have seen that the diode turn-on time is comparable to the diffusion transit time. However, since current through the resistance starts to flow as soon as the source voltage goes positive, the time taken to turn the diode on is not normally a significant parameter. In most practical situations it is the turn-off time that limits a particular circuit behavior. In order to turn the diode off more rapidly, it is common practice to apply a reverse bias voltage of several volts through a resistance. This is the arrangement most often used to characterize the switching behavior of a diode. Figure 3.26 shows the resulting current and junction voltage waveforms as the source voltage is stepped from a forward value of several volts to a reverse value of several volts through a series resistance. The hole distributions at various times during turn-off are shown in Fig. 3.27.

The turn-off transient can now be divided into two distinct phases: (1) a constant reverse current *storage time* t_s, (2) a reverse *decay time constant* t_r. The storage time t_s is defined as the time taken for the diode voltage $v_a(t)$ to decay from its steady-state forward-bias value (of about 0.7 V) to zero. Since the injection law $p_n(0, t) = p_{n0} \exp[v_a(t)/V_t]$, this time is also the time at which $p_n(0)$ reaches the thermal equilibrium value p_{n0}. This is labeled curve 5 in Fig. 3.27. Since the current during the time t_s is set by the reverse source voltage V_r and

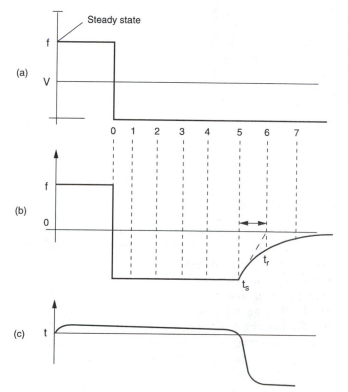

Figure 3.26 (a) Source voltage waveform; (b) diode current waveform; (c) junction voltage waveform $v_a(t)$, for turning off a diode with a reverse voltage source through a series resistance.

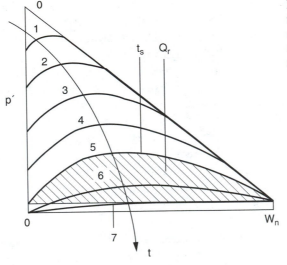

Figure 3.27 Hole distributions versus time during turn-off in the N region of a P^+N narrow-base diode with a reverse source voltage V_r applied through a series resistance R.

the resistance R (assuming that the magnitude of V_r is at least a few volts so that the diode voltage can be neglected during this part of the transient), that is, $I_r = V_r/R$, it follows that the gradient dp/dx is constant at the junction. Toward the right-hand side of the N region, the hole distribution tends to its original steady-state forward value (it takes time for the carriers to adjust their distribution by diffusion). The general shape of the curves in Fig. 3.27 is thus apparent. It is also apparent that at time t_s, a finite amount of hole charge Q_r remains in the N region. The remainder of the transient, the decay phase, corresponds to the evacuation of this residual charge. It may be approximated to an exponential decay with a time constant t_r.

The duration of the constant reverse current storage phase t_s is of particular importance from a circuit point of view. If the current I_r is small in magnitude, it will take a longer time for the voltage to go from about 0.7 V to zero and t_s will be large. If, on the other hand, a large reverse current is used to turn the diode off, t_s will be small. A rough empirical formula relating t_s to the forward-to-reverse current ratio is:

$$t_s = t_b(I_f/I_r)/(1 + I_r/I_f) \tag{3.59}$$

PIN diode switching behavior

In the *PIN* diode the ratio Q/I of steady-state free-carrier charge to dc current is equal to the carrier lifetime τ. It is thus not too surprising that the constant reverse current storage time t_s for the *PIN* diode is determined by the lifetime and not by the diffusion transit time. Figure 3.28 shows the way in which the free carrier charge in a *PIN* diode builds up during turn-on and is removed during turn-off using a voltage source in series with a resistance. t_c is the time constant related to the diffusion transit time, which in this case is due to both holes and electrons diffusing from each side and is given by $t_c = W_i^2/8D$, where D is called the *ambipolar diffusion constant* (a function of both D_n and D_p). More important, it may be shown that the storage time t_s is given to a very good approximation by:

$$t_s = \tau \, \ln(1 + I_f/I_r) \tag{3.60}$$

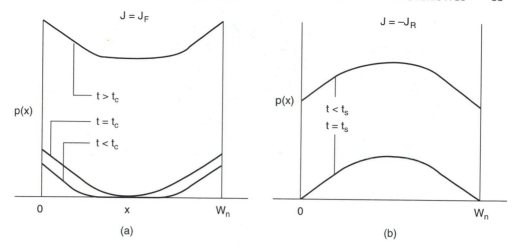

Figure 3.28 Free-carrier distributions in a *PIN* diode during (a) turn-on; (b) turn-off in a circuit with a voltage source in series with a resistance.

This result may be obtained by solution of the charge continuity equation (2.87). It also applies approximately to switching in the wide-base diode.

3.10 TEMPERATURE DEPENDENCE OF DIODE dc CHARACTERISTICS

We can write the general diode *I–V* characteristic from Eqs. (3.11), (3.18), (3.23), and (3.25) as follows:

$$I = I_0' \exp(V_a/mV_t) \qquad (3.61)$$

where $V_t = kT/q$. There are two cases to consider: (1) for a diode operating under low-level injection conditions, but at a high enough bias that space-charge recombination can be neglected [Eq. (3.11)], $m = 1$ and $I_0' \propto n_i^2 \propto \exp(-E_g/kT)$; (2) for a diode operating at either very low or very high bias [Eq. (3.36) or (3.18)], $m = 2$ and $I_0' \propto n_i \propto \exp(-E_g/2kT)$. In general we can hence write:

$$I = C_1 \exp(-E_g/mkT) \exp(qV_a/mkT) \propto \exp[(-1/mkT)(E_g - qV_a)] \qquad (3.62)$$

where for case (1), $m = 1$ and for case (2) $m = 2$. The easist way to examine the current dependence with temperature is to calculate by how much I increases for a small increase in T.

Taking $E_g = 1.1$, $kT(300\text{ K}) = 0.0259$ eV, $V_a = 0.7$ V gives the following results for a 10°C rise in temperature:

1. $m = 1$; current increases by a factor of 1.7 at $V_a = 0.7$ V, or a factor of 2.6 at $V_a = 0.3$ V;
2. $m = 2$; current increases by a factor of 1.3 at $V_a = 0.7$ V, or a factor of 1.7 at $V_a = 0.3$ V.

We can thus conclude that on average the current will appproximately double for a 10°C rise in temperature around room temperature. It is clearly undesirable in a circuit to establish a voltage source across a diode, because the current will be quite unpredictable.

To examine the dependence of voltage on temperature for a fixed current, we can rearrange Eq. (3.62) as:

$$- E_g/mkT + qV_a/mkT = C_2 = \text{const} \tag{3.63}$$

Differentiating with respect to temperature gives:

$$\frac{E_g}{mkT^2} - \frac{qV_a}{mkT^2} + \frac{q}{mkT}\frac{dV_a}{dT} = 0 \tag{3.64}$$

hence:

$$\frac{dV_a}{dT} = \frac{1}{m}\left(\frac{-E_g}{qT} + \frac{V_a}{T}\right) \tag{3.65}$$

At 300 K this gives:
1. For $m = 1$; -2.6 mV/degree at $V_a = 0.3$ V, to -1.3 mV/degree at $V_a = 0.7$ V;
2. For $m = 2$, these values are reduced by one-half.

Thus depending on the voltage bias and operating region (value of m), the diode voltage decreases by between about 0.5 and 2.6 mV per °C at room temperature. This is a very important result, since in most circuit applications the current is held approximately constant in a forward-biased junction.

3.11 METAL–SEMICONDUCTOR CONTACTS AND THE SCHOTTKY DIODE

The built-in barrier potential cannot be "measured" at the terminals of a diode. If this were not the case, then the diode would be a source of power. In order to understand what happens between the diode and the outside world, we need to consider the energy-band system of a metal and a semiconductor and what happens when the two materials are brought together as shown in Fig 3.29.

Note on notation: Some textbooks express work functions and electron affinity in electron volts (eV), using the same symbols ϕ_{ms}, ϕ_F, χ_s, etc. Throughout, we express these quantities in volts.

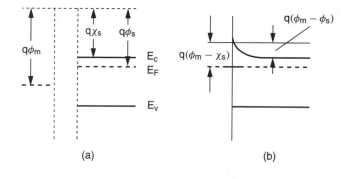

Figure 3.29 Energy-band diagrams for a metal–semiconductor contact with $\phi_m > \chi_s$: (a) Separate system with metal, work function ϕ_m, and semiconductor, work function ϕ_s, electron affinity χ_s; (b) combined metal–semiconductor junction.

The metal is characterized by its work function $q\phi_m$, which is the energy required to completely remove an electron from its surface. In a metal the Fermi level E_F is in the conduction band with nearly all the electrons filling the region between E_F and E_c. In a semiconductor the electron affinity is the energy required to remove an electron from the conduction-band edge, with the work function $q\phi_s$ being defined with reference to the Fermi level as in a metal.

When the two materials are brought into contact, the Fermi levels align, and in the example shown in Fig. 3.29 for the normal situation where $\phi_m > \phi_s$ electrons spill over from the semiconductor to the metal, thus leaving behind a positive charge in the metal. The band edges therefore bend upwards as shown. A barrier for electrons ϕ_b exists analogous to the barrier in a *PN* junction. In this case the barrier is:

$$\phi_b = \phi_m - \chi_s \tag{3.66}$$

This metal–semiconductor barrier is known as a *Schottky barrier*, and the diode thus formed is called a *Schottky diode*. For *P*-type semiconductors, the same argument applies, but the band diagram is modified since the Fermi level is now closer to the valence band E_v. Depending on the doping of the semiconductor and the value of $\phi_m - \phi_s$, the band edges may bend upwards or downwards. Typical values are: Si, $\chi_s = 4.01$ V; GaAs, $\chi_s = 4.07$ V; Al, $\phi_m = 4.28$ V; Pt, $\phi_m = 5.65$ V; W, $\phi_m = 4.55$ V.

Figure 3.30 shows a metal–*P*–*N*–metal system, following the above discussion. In thermal equilibrium the Fermi level E_F must be the same everywhere.

Clearly, under the above conditions a metal–semiconductor barrier is created at either

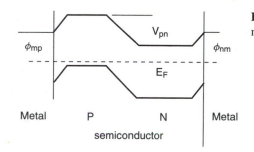

Figure 3.30 Simplified energy-band diagram for a metal–*P*–*N*–metal system.

side of the diode, in such a way that the sum of all three barriers adds up to zero. In the above diagram,

$$\phi_{mp} + V_{pn} + \phi_{nm} = 0 \qquad (3.67)$$

Thus no current flows in thermal equilibrium when the two metal contacts are connected; no external barrier voltage exists.

If the diode had "good" ohmic contacts at either end, all the applied voltage V_a appears across the *PN* junction. In order to create a good ohmic contact, the semiconductor must be heavily doped (N^+ near the contact on the N side and P^+ near the contact on the P side). This will introduce two further small barriers, not shown on the diagram. In the case of the metal to N-type semiconductor, with reference to Fig. 3.29, if the donor concentration is very high, the depletion region (over which the band bending occurs) is very narrow (see following discussion) and current flows by tunneling as in a heavily doped P^+N^+ junction. Contact resistances as low as 10^{-6} ohm cm^{-2} are necessary in high-performance semiconductor devices. This is typical of the values attainable for metal-to-semiconductor junctions with doping levels in excess of 10^{20} cm^{-3}.

Metal–semiconductor junctions also exist that are rectifying (nonohmic). Such a case is the Schottky diode, which is formed by contacting a moderately doped N region with Al. The Schottky diode has an *I–V* characteristic similar to a *P–N* junction, but current flow is mainly due to majority carriers via thermionic emission, with the current density given by:

$$J = A^* T^2 [\exp(-q\phi_b/kT)][\exp(V_a/V_t) - 1] \qquad (3.68)$$

where A^* is the effective Richardson constant ($=120$ A/K^2/cm^2) and ϕ_b is the Schottky barrier height. Values for ϕ_b are typically considerably less than for a *P–N* junction, and the resulting forward-bias current density is several orders of magnitude higher. This means that for a given diode area and diode current, the forward voltage drop (the forward "knee voltage" or "threshold voltage") is a few tenths of a volt less than for a *PN* diode, as illustrated in Fig. 3.31.

In reverse bias the current is correspondingly much higher than for a typical *PN* junction diode.

Since the depletion layer extends only into the semiconductor material, the diode depletion layer thickness d_n is given by the one-sided abrupt junction formula (3.46) with a barrier height ψ_b being the difference between the Fermi levels in the metal and the semiconductor (Fig. 3.29).

$$\psi_b = \phi_m - \phi_s \qquad (3.69)$$

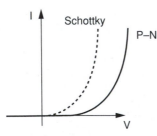

Figure 3.31 *I–V* characteristics of typical *P–N* and Schottky diodes for similar areas.

Note that the position of the Fermi level as a function of doping (required in order to calculate ϕ_s from χ_s) can be obtained from Eq. (F.20). Thus we have:

$$d_n = [2\epsilon(\psi_b - V_a)/(qN_D)]^{1/2} \tag{3.70}$$

and the capacitance is then as in Eq. (3.47):

$$C_j = \epsilon A/[2\epsilon(\psi_b - V_a)/(qN_D)]^{1/2} \tag{3.71}$$

In practical Schottky diodes the surface-state charge will affect the band bending and the values of the above quantities. The surface charge is process dependent and therefore difficult to predict before fabrication.

Example: An Al contact is made to N-type silicon-doped 10^{16} cm^{-3} over an area of $2 \times 2 \ \mu m^2$. Find the threshold voltage for a current of 1 mA and the depletion layer capacitance for a reverse bias of 5 V.

The barrier to electrons, using the data given above and Eq. (3.66), is: $\phi_b = 4.28 - 4.01 = 0.27$ V. From Eq. (3.68), for an area of $2 \times 2 \times 10^{-8}$ cm^2, we have

$$I_0 = 4 \times 10^{-8} \times 120 \times (300)^2 \exp(-0.27/0.0259) = 13 \ \mu A$$

For $I = 1$ mA, we have the corresponding voltage $V_a = V_t \ln(10^{-3}/13 \times 10^{-6}) = 0.112$ V.

Equation (F.20) gives the Fermi-level position above the intrinsic level:
$E_F = E_i + 0.0259 \ln(10^{16}/1.6 \times 10^{10}) = E_i + 0.346$ eV. Since $E_g = 1.12$ eV,
$E_c - E_F = 0.5 \times 1.12 - 0.346 = 0.214$ eV. The value of ϕ_s (see Fig. 3.29) is thus
$\chi_s + 0.214 = 4.01 + 0.214 = 4.224$ V. The junction barrier potential is thus:
$\psi_b = 4.28 - 4.224 = 0.056$ V. The junction capacitance at -5 V is given by Eq. (3.71) as:
$C_j = 10^{-12} \times 4 \times 10^{-8}/[2 \times 10^{-12}(0.056+5.0)/(1.6 \times 10^{-19} \times 1 \times 10^{16})]^{1/2} = 4.8 \times 10^{-16} = 0.48$ fF.

It is important to note that since only majority carrier current flow is involved in the Schottky-diode thermionic emission process, there is no minority carrier diffusion capacitance. The speed of the Schottky diode is therefore limited only by the above depletion-layer capacitance, and the diode is ideal for high-speed switching. Because of this speed advantage combined with the low "knee" voltage, Schottky diodes are widely used for clamping applications (Schottky TTL and STL logic families).

3.12 CONCLUSIONS

In this chapter we started by introducing the reader to the narrow-base diode, one of several diode structures, particularly important because of its relevance to the bipolar transistor to be studied in Chapter 4. The dc characteristics of the high-current *PIN* diode were then derived, followed by the treatment of high-level injection in the narrow-base diode. The wide-base, P^+NN^+, and narrow-base diode with recombination were reviewed.

Low forward-bias behavior due to space-charge recombination was then presented. This is common to all diode types. The reverse-bias analysis covered first space-charge generation, and then avalanche multiplication and breakdown. A summary of the overall current–voltage characteristics was then given.

The small-signal depletion layer and diffusion capacitances were then derived, leading to the small-signal equivalent circuit of the diode for high-frequency modeling. A qualitative discussion of switching transients in diodes was given, with emphasis on "current switching." A brief treatment of the temperature dependence of diode I–V characteristics was given, and the chapter concluded with an overview of metal–semiconductor junctions and the Schottky diode.

PROBLEMS

1. Consider a P^+N narrow-base diode with $N_A = 10^{19}$ cm^{-3}, $N_D = 10^{15}$ cm^{-3}, $W_p = 1\,\mu$m, $W_n = 30\,\mu$m, cross-sectional area 1 cm^2. Calculate the following, for a forward bias $V_a = 0.5$ V:

   a. The depletion-layer thickness on the P and N sides;

   b. The hole and electron currents and the terminal current;

   c. The ideal reverse saturation current I_0.

2. For the diode of Problem 1 calculate:

   a. The injected carrier concentrations on each side of the depletion layer;

   b. The voltage at which high-level injection starts.

3. For the diode of Problem 1, calculate the hole diffusion length for lifetimes of (a) $10\,\mu$s; (2) $0.1\,\mu$s, and sketch $p(x)$ for these two cases. Is the narrow-base diode a valid assumption in each case? Calculate the recombination current and the total current for both these carrier lifetimes, for the bias conditions of Problem 1.

4. A *PIN* diode has a center-layer thickness $W_i = 50\,\mu$m and a carrier lifetime $\tau_i = 10\,\mu$s. If the cross-sectional area is 1 cm^2, calculate the terminal current for $V_a = 0.8$ V. Calculate the carrier diffusion length (use $D_i = 10$ cm^2/s) and sketch the carrier distribution in the center layer. Would the *PIN* diode theory be valid if $\tau_i = 0.1\,\mu$s?

5. Consider a P^+N narrow-base diode operating under high-level injection on the N side. By setting the electron current equal to zero, show that the N-side neutral-region voltage drop is given by: $V_{an} = V_t \ln[p_n(0)/N_D]$. If we assume that the injection law is still valid for the fraction V_{aj} of the applied voltage that occurs across the space-charge region [i.e., $p_n(0) = p_{n0} \exp(V_{aj}/V_t)$], show that $p_n(0) = n_i \exp(V_a/2V_t)$, where $V_a = V_{aj} + V_{an} = $ total terminal bias. Since the current is proportional to $p_n(0)$, this illustrates the $V_a/2V_t$ current dependence under HLI conditions.

6. Consider a P^+NN^+ diode with $N_A^+ = N_D^+ = 10^{18}$ cm^{-3}, $N_{Depi} = 10^{15}$ cm^{-3}, $W_n = 20\,\mu$m, $\tau = 1\,\mu$s, diameter 100 μm, series contact resistance $= 100$ ohm. Calculate the following:

   a. Ideal I_0;

   b. I_{0r} due to space-charge recombination;

   c. V_{aHLI} at which high-level injection starts;

d. The value of V_a at which space-charge recombination current equals the (ideal) normal neutral-region current;

e. The reverse leakage current due to space-charge recombination at $V_a = -10$ V;

f. The avalanche breakdown voltage.

Sketch the forward $I–V$ characteristic on a semilog plot for a terminal voltage from 0 to 1 V.

7. Show that for a P^+N narrow-base diode the ratio R_{NSCR} of reverse neutral-region diffusion current to reverse space-charge generation current is given by $R_{NSCR} = (n_i/N_D)(D_p\tau)/(W_n d_n)$. Calculate this ratio for: (a) a silicon diode with the values used in Problem 1 with $\tau = 1\ \mu s$; (b) a germanium diode with the same parameters [use Eq. (2.2) at $T = 27°C$ and calculate n_i for Ge, whose band gap is 0.66 eV compared to silicon, 1.12 eV, assuming the same values of N_c, N_V].

8. For the diode of Problem 1 calculate the magnitude of the electron concentration required in the space-charge layer to support the ideal current I_0 (assuming very high lifetime). *Hint:* Assume high-field drift current for current transport in the space-charge layer.

9. For the diode of Problem 1 calculate and *sketch* (use linear–linear axes) the depletion-layer capacitance and the diffusion capacitance from $V_a = -10$ to $+0.8$ V. Calculate the forward voltage at which the two capacitances have equal values.

10. Repeat the calculations of Problem 9 if the diode is the *PIN* diode of Problem 4 if $N_D = 10^{14} \text{cm}^{-3}$.

11. Calculate the reverse breakdown voltages for the following *PN* diodes (assume the *P*-side doping is 10^{21} cm^{-3} in all cases): *N*-side doping: 10^{15}, 10^{17}, 10^{19}, 10^{21} cm^{-3}. Assume $E_{BR} = 3 \times 10^5$ V/cm in each case. Discuss briefly the last two cases.

12. Use Eq. (F.20) and sketch the band diagrams for the junctions used in Problem 11.

13. Derive the result for depletion-layer capacitance Eq. (3.45) using Eqs. (2.61) and (3.44).

14. The following capacitance measurements were obtained on a diode with a cross-sectional area of 10^{-5} cm^2:

C(pF)	0.37	0.58	0.72	0.87	1.2
V_a	−2.0	−0.4	0.0	0.30	0.54

Use a plot of $1/C^2$ versus V_a to determine the value of built-in barrier voltage V_{bi}. Use this result with Eq. (3.47) to determine also the doping level of the lightly doped side.

15. For the diode of Problem 1, calculate the base transit time and the small-signal equivalent circuit parameters for: (a) $I = 1\ \mu A$; (b) $I = 1$ mA.; (c) $I = 50$ mA.

16. The narrow-base diode of Problem 1 is used in the circuit of Fig. 3.25 with a circuit resistance of 1 kohm and a voltage going from −5 to +5 V at time $t = 0$, then back to −5 V after 100 μs. Calculate the constant reverse current storage time t_s. Calculate the zero-bias capacitance. Assume this to be the average capacitance when the diode voltage is going from −5 to +0.5 V and use circuit theory to estimate the time for the diode voltage to become positive. *Hint:* Use the result of Eq. (4.93).

17. Repeat Problem 16 for the *PIN* diode of Problem 4. Assume a doping level of 10^{15} cm^{-3}.

18. A silicon diode has $I_0 = 10^{-15}$ A, $m = 1$, at 27°C. It is biased by a voltage source $V_a = 0.6$ V. Calculate the factor by which the current is increased if the temperature increases to 40°C. If the

same diode is biased from a voltage of 10 V in series with a 1 kohm resistance, calculate the decrease in forward-diode voltage for the above temperature change.

19. A silicon Schottky diode is made with a Tungsten contact on *N*-type material doped 10^{16} cm^{-3}. The dimensions are $10 \times 10 \, \mu$m^2. Calculate the forward bias at a current of 1 mA. Compare this with the voltage for a P^+N narrow-base diode on the same material of thickness $W_n = 10 \, \mu$m. Calculate the depletion-layer capacitance of both diodes at zero bias and the RC time constant for a circuit resistance of 1 kohm. Compare this to the t_s of the narrow-base diode for $I_F/I_R = 1$.

20. Design a silicon narrow-base diode for a breakdown voltage of 50 V. What is the value of the constant reverse current storage time for $I_F/I_R = 1$. Calculate the area required for a current of 1 A at $V_a = 0.7$ V and hence the value of zero-bias junction capacitance.

21. Repeat Problem 20 for a *PIN* diode and estimate approximately the minimum value of carrier lifetime required for true *PIN* performance.

22. For a linearly graded junction, use Eq. (2.67) to show that the capacitance–voltage law is given by: $C(V) = C_0/[1 - (V_a/V_{bi})]^{1/3}$. Derive the expression for C_0 for (a) the linearly graded junction; (b) the one-sided abrupt junction.

References

1. S. M. Sze, *Physics of Semiconductor Devices*. 2nd ed., New York: Wiley-Interscience, 1981.
2. E. S. Yang, *Fundamentals of Semiconductor Devices*. New York: McGraw–Hill, 1988.
3. A. Bar-Lev, *Semiconductors and Electronic Devices*. 3rd Ed., New York: Prentice-Hall, 1993.
4. D. H. Navon, *Semiconductor Microdevices and Materials*. New York: Holt, Rinehart, Winston, 1986.
5. M. S. Tyagi, *Introduction to Semiconductor Materials and Devices*. New York: Wiley, 1991.
6. P. E. Gray, D. DeWitt, A. R. Boothroyd, and J. F. Gibbons, *Physical Electronics and Circuit Models of Transistors*. Semiconductor Electronics Education Committee, Vol. 2, New York: Wiley, 1964.
7. D. J. Roulston, *Bipolar Semiconductor Devices*. New York: McGraw–Hill, 1990.
8. D. A. Neamen, *Semiconductor Physics and Devices*. Chicago: Irwin, 1997.

Chapter 4

Bipolar Junction Transistors

4.1 INTRODUCTION

The original bipolar transistor invented in 1948 was made of germanium and was a "point contact" structure. It was subsequently found that silicon technology using diffused junctions offered several advantages from an industrial point of view. One of the main attractions of silicon is the ease with which SiO_2 (essentially the same as glass) can be grown on top of a silicon wafer, thus providing not only an ideal barrier (defined by a geometrical mask pattern) to dopant impurity atoms, but also a near-perfect insulating layer on top of which metal lines can be deposited for contacts and, in the case of integrated circuits, interconnects. Silicon "planar" technology (as originated simultaneously at Texas Instruments and Fairchild Laboratories in 1959) is universally used today in fabricating the BJT both for integrated circuits and for a wide range of discrete transistor structures. The size of an individual BJT varies from the low-voltage sub-micron VLSI device to large high-voltage (600 V) power-switching structures more than 1 cm in diameter. Technological advances have enabled huge performance improvements in recent years; techniques include polysilicon emitters and heterojunction bipolar transistors using GaAlAs/GaAs or Si/SiGe junctions. The most recent SiGe devices have unity current gain frequencies (f_t) of order 100 GHz.

In this chapter we describe briefly the vertical *NPN* BJT structure with a double diffused impurity profile. We then focus on a simplified "constant doping" impurity profile, which lends itself to simple analysis and is adequate even for "roughing out" engineering designs, providing certain precautions are taken. The analysis starts by considering electron and hole currents to build up a picture of the dc characteristics of the BJT. High-current effects causing performance degradation and breakdown voltage are then studied. Small-signal behavior is then discussed, with capacitance terms added as was done for the diode in Chapter 3; this leads to the result for the high-frequency f_t and f_{maxosc} figures of merit. The integrated circuit BJT is presented, with an outline of the properties of the lateral *PNP* structure. We then present switching properties and temperature effects. This is followed by a short section on BJT noise performance and an overview of the basic concepts of the Gummel–Poon SPICE CAD model. The chapter concludes with a summary of limitations of the theory used in the previous sections.

4.2 THE VERTICAL *NPN* BIPOLAR TRANSISTOR STRUCTURE

Figure 4.1 shows the essential features of the discrete vertical *NPN* bipolar transistor. The starting material is moderately doped *N*-type silicon on a heavily doped N^+ substrate. Figure 4.1(a) shows the rectangular mask areas through which the two successive diffusions are made. Boron (*P*-type impurities) is diffused at high temperature (of order 1000°C) through the large rectangular area to create a P^+N junction in a manner identical to the fabrication of a diode as discussed in Chapter 3; this is the base diffusion. A second diffusion, this time using phosphorus or arsenic (*N*-type impurities) is then made through the central rectangle to form the emitter; the innermost rectangle is the metal contact area to the emitter; the two small rectangles on either side are the base contacts. Contact to the collector is made at the bottom of the wafer to the N^+ region. The horizontal dimensions of the emitter range from a few tenths of a micrometer square for high-speed VLSI devices to over 100 μm for the shortest dimension in high-current high-voltage BJTs. This fabrication sequence produces a structure with two *N* regions (top and bottom) and a *P* region sandwiched between them. Because there are now two junctions (two overlapping diodes with a common *P* region), there are two space-charge regions (depletion layers), shown shaded. The SiO_2 oxide layer, also shown shaded, is used as the barrier to dopants; this is discussed in Chapter 8. Base–collector junction X_{jbc} depths range from 10 μm for high-voltage devices to less than 0.1 μm for small high-speed transistors. Emitter–base junction depths X_{jeb} are typically about one-half of X_{jbc}. A crucial property of the fabrication is the distribution of donor and acceptor atoms. Figure 4.2 shows the same cross section with the impurity profile of the transistor beside it. This is normally drawn for convenience with the depth x as the horizontal axis, as shown in Fig. 4.3. Industrial design of the bipolar transistor hinges around a precise study of how electron and hole current flows in each of the three regions. This is invariably done using numerical computer simulation; such an analysis takes into account such factors as the electric field produced by the nonuniform impurity profile, and the variation of mobility and carrier lifetime with doping level. Furthermore, the boundaries

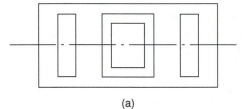

(a)

Figure 4.1 Bipolar transistor structure: (a) top view (mask diagram); (b) cross-sectional view (space-charge layers and SiO_2 regions are shown shaded). The emitter is *N* type, the base is *P* type, and the collector is *N* type.

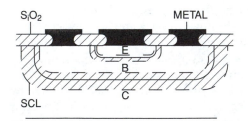

(b)

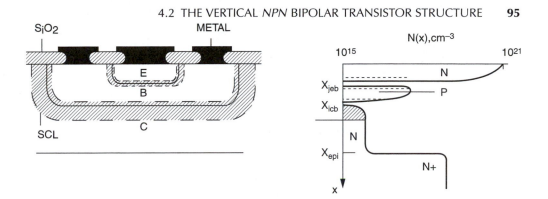

Figure 4.2 BJT cross section with impurity profile, aligned on same vertical axis.

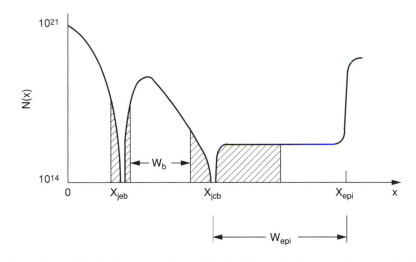

Figure 4.3 Impurity profile of a bipolar transistor with depth from the silicon surface as the horizontal *x* axis. (See also CD-ROM BIPGRAPH example TRAN1; "Net doping vs depth.")

of the two space-charge layers are not trivial to determine and depend on both current density and bias voltage. Only the uniformly doped collector region can be analyzed with a useful degree of accuracy with simple formulae. In fact, it is this region that forms the starting point of a design; the doping level N_{epi} and thickness of the moderately doped epitaxial layer (see Section 8.2.2) W_{epi} are chosen to provide the required breakdown voltage.

It is clearly difficult to study electron and hole currents with a doping level that varies with depth. Only the uniformly doped collector region can be studied accurately with simple analytic formulae. It is common practice (and is of considerable value from a design viewpoint) to replace this impurity profile with constant doping in each region, as shown in Fig. 4.4. This can then be studied using the approach used for diodes in Chapter 3.

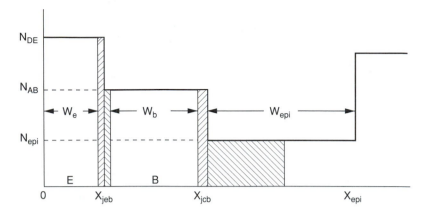

Figure 4.4 Simplified impurity profile of BJT with constant doping in each region.

The analysis will be carried out assuming the normal bias arrangement of a forward-biased emitter–base (e–b) junction and a reverse-biased collector–base (c–b) junction. This is one reason why the e–b space-charge layer in Figs. 4.3 and 4.4 is shown as being much narrower than the c–b space-charge layer. The second reason is that the collector is moderately doped compared to the base and emitter regions; from Chapter 2, we recall that the depletion layer is wider for light doping. The bias is most frequently arranged as shown in Fig. 4.5 (in simplified form), which is the common-emitter configuration. The forward bias V_{BE} between base and emitter is normally set up through a resistance. There is also a load resistance R_L between the collector terminals and several volts of collector supply voltage V_{cc}. Since V_{BE} is only about 0.7 V, a supply voltage V_{cc} of a few volts is adequate to ensure that the collector–base junction is reverse biased, even allowing for a voltage drop across R_L.

The collector–base V_{BE} and collector emitter V_{CE} voltage are related by:

$$V_{CB} = V_{CE} - V_{BE} \tag{4.1}$$

4.3 dc ANALYSIS OF THE *NPN* BIPOLAR TRANSISTOR

Using the bias described in Section 4.2, Fig. 4.6 shows the electron and hole distributions in the base, collector, and emitter regions. The forward V_{BE} voltage creates injected electron and hole concentrations at the e–b depletion layer boundaries, given by the junction injection law, Eqs. (2.74) and (2.75):

$$n_b(0) = n_{b0} \exp(V_{BE}/V_t) \tag{4.2}$$

$$p_{ne}(0) = p_{ne0} \exp(V_{BE}/V_t) \tag{4.3}$$

We have used the subscript b for the P-type base region, and the subscript ne for the n-type emitter region (to distinguish from the N-type collector region).

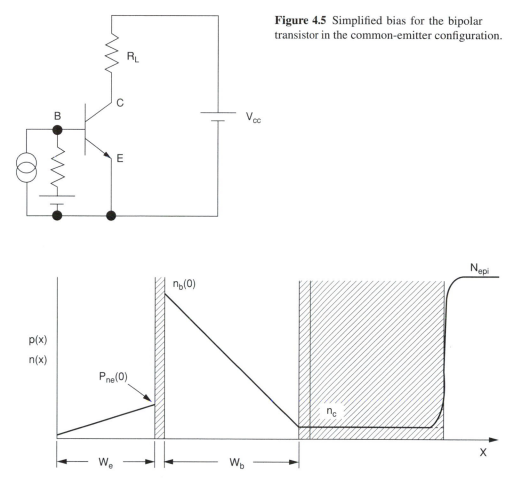

Figure 4.5 Simplified bias for the bipolar transistor in the common-emitter configuration.

Figure 4.6 Electron and hole distributions in a normally biased BJT. Space-charge layers are shown shaded. (See also CD-ROM BIPGRAPH example TRAN2; "Neutral base minority carrier conc. vs depth" and "Collector-base SCL $n(x)$"; see also "Vertical results for all x.")

Applying the injection law at the collector side of the base region gives:

$$n_b(W_b) = n_{b0} \exp(V_{BC}/V_t) \tag{4.4}$$

Note that under normal bias V_{CB} is positive and V_{BC} is negative (following conventional use of subscripts). This gives the result that $n_b(W_b) = 0$. It should be recalled, however, that the injection law was derived from the Boltzmann equations, strictly only valid in thermal equilibrium. It turns out that the result is not quite accurate in the case of this b–c junction. In fact, for electron current to flow in the base–collector space-charge layer, a finite electron concentration n_c must exist, as shown in Fig. 4.6. It is a simple matter to estimate the value of n_c, since the electric field is fairly high in this space-charge region (a few volts within about 1 μm gives a field in excess of 10^4 V/cm), and it can safely be assumed that the electrons are drifting at their saturated drift velocity $v_s \approx 10^7$ cm/s (see Fig. 2.10). The electron current I_{nC} in the b–c space charge layer is thus (2.19):

$$I_{nC} = q A v_s n_c \tag{4.5}$$

Clearly, n_c will be quite small, and we shall see that for most transistors it can safely be set to zero in calculating the gradient of the electron concentration in the base region. However, it must be remembered that a finite (albeit small) value of n_c is essential for current to flow through the space-charge region.

The currents flowing due to diffusion in the base and emitter regions are given by the formulae derived in Chapter 3 for the narrow-base diode with the area A_E being that of the emitter (given the very small vertical x dimensions of order a micron or less, we will assume for the moment that the carrier diffusion lengths are much greater than the region thicknesses):

$$I_{nC} = q A_E D_{nb} \frac{dn}{dx} \tag{4.6}$$

$$= -q A_E D_{nb} n_b(0) / W_b \tag{4.7}$$

$$I_{pE} = -q A_E D_{pe} \frac{dp}{dx} \tag{4.8}$$

$$= -q A_E D_{pe} p_{ne}(0) / W_e \tag{4.9}$$

We have added subscripts to diffusion coefficients to denote D_{nb} for D_n in the base and D_{pe} for D_p in the emitter. We assume $V_{BE} \gg V_t$ (certainly true for V_{BE} of order 0.7 V) such that we can neglect the p_{ne0} term in calculating the hole concentration gradient.

The electrons diffusing across the base are swept across the high-field b–c space-charge layer and then flow to the collector contact in the N-doped region as low-field drift current. Note that although I_{nC} is negative (electrons flowing from left to right in Fig. 4.6), the collector current (defined conventionally as positive flowing from the positive V_{cc} supply into the collector terminal) is positive.

$$I_C = -I_{nC} \tag{4.10}$$

The holes injected into the emitter clearly flow from right to left in Fig. 4.6, thereby giving a negative component of conventional current at the emitter terminal. The electron current becomes majority carrier (low-field) drift current in the N-type emitter and simply adds to the hole current. The total emitter current is thus:

$$I_E = I_{nC} + I_{pE} \tag{4.11}$$

It may be noted that *the collector current is governed by the properties of the base region and the base current is governed by the properties of the emitter region*. The question now arises: Where does the hole current flow from in the base region? To answer this question, we must look at the second dimension of base region of the BJT. This is shown in Fig. 4.7.

As already discussed above, the emitter current is the sum of the collector current and of the hole current injected into the emitter I_{pE}. This hole current is injected from the P-type base

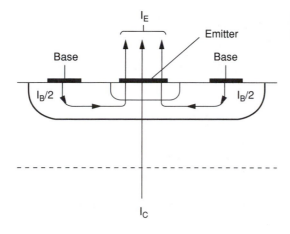

Figure 4.7 Components of current in a normally biased BJT.

region and must come from the base contacts on either side of the emitter as shown. Thus we have a base current I_B given by:

$$I_B = -I_{pE} \tag{4.12}$$

$$= q A D_{pe} p_{ne}(0) / W_e \tag{4.13}$$

This is positive, that is, conventional current flowing into the base terminal.

We are now in a position to calculate one of the most important parameters of the BJT, the common emitter-current gain h_{FE} or β defined as the ratio of collector to base current:

$$\beta = I_C / I_B = [q A D_{nb} n_b(0) / W_b] / [q A D_{pe} p_{ne}(0) / W_e] \tag{4.14}$$

Substituting for the injected carrier concentrations from the injection law gives:

$$\beta = (D_{nb} n_{b0} W_e) / (D_{pe} p_{ne0} W_b) \tag{4.15}$$

Rearranging and using the mass action law: $n_{b0} = n_i^2 / N_{AB}$; $p_{ne0} = n_i^2 / N_{DE}$, where N_{AB} is the acceptor base doping level and N_{DE} is the emitter donor doping level, gives:

$$\beta = (D_{nb} N_{DE} W_e) / (D_{pe} N_{AB} W_b) \tag{4.16}$$

At first sight, it would appear that is a simple matter to obtain extremely high values of current gain.

Example of gain calculation using above classical formula:

Emitter doping 10^{20} cm^{-3}; base doping 10^{17} cm^{-3}; emitter and base thickness $W_e = W_b = 1.0$ μm:

For an emitter doping level of 10^{20} cm^{-3}, Fig. 2.11 gives the hole mobility $\mu_p \approx 50$ cm^2/V s (by extrapolation); hence from the Einstein relation $D_{pe} = V_t \mu_p = 1.25$ cm^2/s.

For a base doping of 10^{17} cm^{-3}, Fig. 2.11 gives $\mu_n = 800$ cm^2/V s, hence $D_{nb} = 0.025 \times 800 = 20$ cm^2/s. We thus find

$$\beta = 20 \times 10^{20} \times 1.0 \times 10^{-4}/(1.25 \times 10^{17} \times 1.0 \times 10^{-4}) = 1.6 \times 10^4$$

However, in Chapter 2, we saw that for doping levels in excess of about 10^{19} cm^{-3}, the band-gap narrowing (BGN) phenomenon actually reduces the effective doping level for minority carrier transport calculations. Figure 4.8 shows a plot of real N_D and effective (N_{Deff}) doping in the emitter of a BJT. It can be seen clearly from this plot that in practice the effective emitter doping level is limited to not much greater than 10^{18} cm^{-3}; so a more useful and much more accurate equation for current gain is as follows:

$$\beta = (D_{nb}/D_{pe})(N_{Deff}/N_{AB})(W_e/W_b) \tag{4.17}$$

where $N_{Deff} \approx 10^{18}$ cm^{-3} at room temperature.

The above current gain is reduced by a factor $10^{18}/10^{20} = 1/100$ giving a final value $\beta = 160$.

The current due to recombination of minority carrier holes in the emitter can be determined approximately using the narrow-base diode approach, Eq. (3.28). This additional component of base current can be expressed from Eq. (3.28) as $I_{Erec} = Q_h/\tau_{he}$, where τ_{he} is the hole lifetime in the emitter. The ratio of hole current without recombination I_{pE}, to I_{Erec} is given by Eq. (3.31a) as

$$I_{pE}/I_{Erec} = 2(L_{pe}/W_e)^2 \tag{4.18}$$

where L_{pe} is the hole diffusion length in the emitter. For high doping levels the carrier lifetime can be 1 ns or less; so L_{pe} is of order 0.3 μm or less. Clearly, for emitter junction depths comparable to or greater than this value, recombination in the emitter will have a considerable effect on gain calculations. For $W_e \leq L_{pe}$ we can combine Eqs. (4.17) and (4.16) to give the *current gain in the presence of emitter recombination* as:

$$\beta_{erec} = (D_{nb}/D_{pe})(10^{18}/N_{AB})(W_e/W_b)/[1 + 0.5(W_e/L_{pe})^2] \tag{4.19}$$

Note that we have implicitly neglected recombination in the neutral base. In Section 3.5.3 we saw that the recombination current due to the triangular distribution of charge Q recombining with a lifetime τ gives a recombination current $I_{rec} = Q/\tau$. In the base of the BJT the triangular distribution gradient corresponds to the collector current I_C, whereas the recombination component Q/τ is supplied by majority carrier current from the base terminal. The ratio I_C/I_{rec} represents a third component of the *current gain due only to base region recombination* β_{brec}, which from Eq. (3.31a) is:

$$\beta_{brec} = 2(L_{nb}/W_b)^2 \tag{4.20}$$

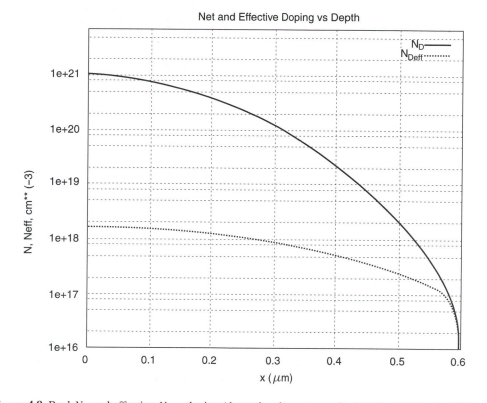

Figure 4.8 Real N_D and effective N_{Deff} doping (due to band-gap narrowing) in the emitter of a BJT at 300 K generated from the BIPOLE3 program [8]. (See also CD-ROM BIPGRAPH example TRAN1 or TRAN2; "Net and effective doping vs depth.")

where L_{nb} is the electron diffusion length in the base. Since typical values of L_{nb} are of order 30 to 100 μm and W_b is of order 0.1 to 1 μm for small transistors, the value of β_{rec} is of order 2×10^3 to 2×10^6 and is so high that it can be neglected in all but deep-base high-voltage transistors ($W_b \gg 1\mu$m).

It is also important to note that the above calculation of hole current injected into the N-type emitter is a gross simplification to the real situation. The diffused doping profile introduces a large electric field, and a strong dependence of mobility with depth; these factors, combined with a carrier lifetime dependence on doping means that hand calculations can give at best only very general trends for the value of the hole current injected into the emitter.

The calculation of collector current is more valid, mainly because the doping level in the base rarely exceeds 10^{18} cm^{-3} and so BGN is not too important. Also, since the base doping level is not too high, mobility and lifetime variation with doping may be neglected to a first approximation. If N_{AB} is taken to be an average base doping level, the above result for collector current is a fair estimate.

Example of modified current gain calculation for above transistor: Assume emitter carrier lifetime = 10 ns, base carrier lifetime = 1 μs.

$$L_{eh} = \sqrt{D\tau_e} = \sqrt{1.25 \times 10 \times 10^{-9}} = 1.1 \times 10^{-4} = 1.1\,\mu\text{m}$$

$$L_{nb} = \sqrt{D\tau_e} = \sqrt{20 \times 10^{-6}} = 4.5 \times 10^{-3} = 45\,\mu\text{m}$$

$1 + 0.5(W_e/L_{pe})^2 = 1.4$, and substituting into Eq. (4.19) gives:

$$\beta_{\text{erec}} = (20/1.25)(10^{18}/10^{17})(10^{-4}/10^{-4})/1.4 = 113$$

Equation (4.19) gives a limit for gain due to base recombination as:

$$\beta_{\text{brec}} = 2(4.5/1)^2 = 4050$$

Clearly this is so high as to have no effect on the final gain. The value of $\beta = 113$ is a very realistic estimate of current gain.

We see from the above that a small value of neutral base width W_b will give a high current gain. However, a small value of base width leads to a large value of base resistance, which is undesirable. The base resistance can be evaluated as follows, with reference to Fig. 4.9. This shows the current entering from both sides of the emitter. Let us denote the emitter width, in the direction of base current flow, by L, and the emitter dimension perpendicular to base current flow by B (the emitter area is thus $A_e = LB$). The thin base region, of thickness W_b, has a resistivity for a given base doping level given by:

$$\rho_b = 1/(q\mu_p N_{AB}) \tag{4.21}$$

The resistance associated with the base region, length L, is:

$$R_B = \rho_b L/(W_b B) \tag{4.22}$$

However, in this discrete transistor, with two symmetrical base contacts, the current comes in from both sides of the emitter and therefore travels only a distance $L/2$. Furthermore, the two base contacts are connected together by the base metallization; this is equivalent to two

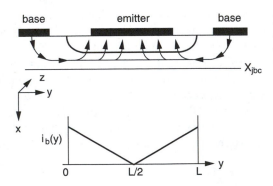

Figure 4.9 Base current flow for base resistance calculation. Arrows represent direction of base current flow. The second diagram shows how the horizontal y component of i_b varies in the y direction.

resistances in parallel. There is thus a reduction factor $\frac{1}{4}$. A third reduction factor arises from the fact that the base current actually reduces in the horizontal direction from the edge to the center, because it is "used up" to supply the hole current injected into the emitter as shown in Fig. 4.9. It can be shown that for purposes of calculating power loss this introduces a further factor of $\frac{1}{3}$. The final value of base resistance seen at the terminals is thus given for a double-base contact transistor by [7]:

$$r_{bb} = \rho_b L / (12 W_b B) \tag{4.23}$$

Before leaving this topic, let us note that the ratio ρ_b / W_b is easily measured from a test pattern and is called the *pinched-base sheet resistance R_{be}/sq*. It is one of the most important measurements made routinely to check a new process in transistor manufacturing. In terms of the doping level we may write:

$$R_{be}/sq = 1/(q \mu_p N_{AB} W_b) \tag{4.24a}$$

and thus:

$$r_{bb} = (R_{be}/sq)(L/12B) \tag{4.24b}$$

The dc characteristics of the BJT from the above simplified analysis are given directly by replacing the injected carrier concentrations $n_b(0)$ and $p_{ne}(0)$ by their voltage dependence. Thus:

$$I_C = I_{cs} \exp(V_{BE}/V_t) \tag{4.25}$$

$$I_B = I_{bs} \exp(V_{BE}/V_t) \tag{4.26}$$

where

$$I_{cs} = q A D_{nb} n_{b0} / W_b \quad \text{and} \quad I_{bs} = q A D_{pe} p_{ne0} / W_e \tag{4.27}$$

and where we have neglected recombination in the base and emitter. Note that these quantities appear to be independent of the collector–base voltage V_{CB}. This is a good approximation, but it may be observed that since the space-charge layer thickness of the c–b junction varies with V_{CB}, clearly there will be some dependence of the neutral base width W_b on V_{CB}. This is a second-order effect, to be discussed briefly later in Section 4.6.

4.4 dc BEHAVIOR

4.4.1 Operating regimes

Saturation regime

So far we have restricted discussion to the *forward or normal active regime* where V_{CB} is positive; that is, the collector–base junction is reverse biased. In some switching applications of the BJT (e.g., TTL logic, high current switching), V_{CE} is forced to a low value by the application of a large base current I_B in Fig. 4.5, thus forcing I_C to a high value and $V_{CB}(= V_{CE} - V_{BE})$ becomes less than zero (I_C is then limited to slightly less than V_{cc}/R_L). This is referred to as the saturation region of operation. The corresponding electron concentration distribution in the base region is shown in Fig. 4.10. The broken curve in the collector neutral region represents holes injected by the forward-biased b–c junction.

The injection law at the b–c junction gives the electron concentration at W_b:

$$n_b(W_b) = n_{b0} \exp(V_{BC}/V_t) \tag{4.28}$$

The concentration gradient in Eq. (4.6) is now reduced, giving a lower value of collector current $I_C = -I_{nC}$:

$$I_C = (qAD_{nb}n_{b0}/W_b)[\exp(V_{BE}/V_t) - \exp(V_{BC}/V_t)] \tag{4.29}$$

Substituting $n_{b0} = n_i^2/N_{AB}$ gives:

$$I_C = [qAD_{nb}n_i^2/(N_{AB}W_b)][\exp(V_{BE}/V_t) - \exp(V_{BC}/V_t)] \tag{4.30}$$

In Appendix O we show that in general the product $N_{AB}W_b$ may be replaced by the integral of the base doping $G_b = \int_0^{W_b} N_A(x)\,dx$. This is called the *Gummel integral* and is a key parameter of the BJT.

It should also be recognized that with positive values of V_{BC}, holes will now be injected into the N-type collector region, as indicated by the broken curve of Figure 4.10. This will create an additional component of base current I_B similar to that injected into the N-type emitter (and a corresponding small change in collector current). The net result is a decrease in current gain I_C/I_B (I_C decreases, I_B increases). This modified gain is referred to as the *forced gain β_f*.

It is instructive to consider how the transistor enters the saturation regime. Figure 4.11(a) shows the electron distribution $n(x)$ in the neutral base region under constant-collector current (I_C) conditions. This is a common situation in a common-emitter circuit, where the base drive I_B is increased with I_C held at a constant value by a load resistance and constant supply voltage. V_{BE} and V_{BC} will increase at the same rate so that the right-hand side of Eq. (4.30) stays constant. In Fig. 4.11(b), the transistor enters saturation with V_{BE} being held constant, hence a fixed value of injected electron concentration $n_b(0)$. As V_{BC} increases in the forward-bias

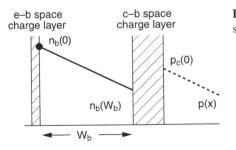

Figure 4.10 Electron distribution in the base region for saturation where both V_{BE} and V_{CB} are forward biased.

direction, the gradient dn/dx is decreased (curve Q), and I_C decreases, eventually changing sign (curve P). It is rare for this condition to be encountered in a practical circuit, but this behavior explains the curve tracer characteristics of Figs. 4.13 and 4.14.

Inverse active regime

Figure 4.12 shows the currents that flow when the transistor is biased with V_{BE} at a reverse voltage and V_{BC} at a forward voltage. In this case the roles of emitter and collector are interchanged. Collector current now flows "upwards" to the N diffused "collector." The emitter hole current density J_{pv} flows vertically under the whole of the base diffusion into the epitaxial "emitter" layer. In addition, electrons flow "upwards" to the base contact and to the oxide surrounding the base contact and the diffused "collector" with current densities J_{em} and J_{eox}. Because of the large areas compared to the diffused "collector," the additional base current due to J_{pv}, J_{em}, J_{eox} means that the "upward" or inverse current gain will be low, typically of order 1 to 5.

A practical use of the upward-operating transistor is in integrated injection logic (IIL). In this technology the upward-operating transistor has two or three collectors and is coupled with a lateral PNP (Section 4.7.2) to form a highly compact lower power logic gate [7].

Cut-off regime

The BJT is said to be in the cut-off region when both V_{BC} and V_{BE} are reverse biased. Ideally the currents will be given by the diode equations (3.12a) for holes in the emitter and (3.26) for holes in the collector (which behaves like the center layer of a P^+NN^+ diode). However, as explained in Chapter 3, at moderate reverse bias, space-charge generation current dominates in silicon over neutral-region diffusion current. Equation (3.39b) can be used to estimate the magnitudes.

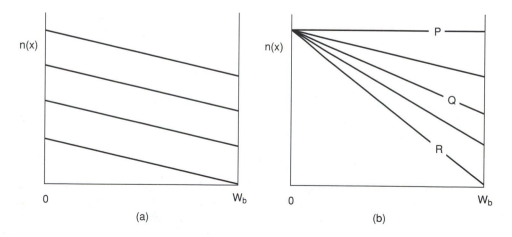

(a)   (b)

Figure 4.11 (a) Electron concentration in the base region as the transistor enters saturation under constant I_C conditions. (b) Electron concentration in the base region as the transistor enters saturation under constant V_{BE} conditions, with V_{BC} increasing.

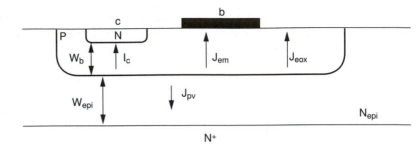

Figure 4.12 Currents for the inverse active regime of operation. The arrows indicate direction of carrier flow.

Example of "ideal" reverse leakage current calculation: Assume the same transistor values as in the previous example in Section 4.3, adding a collector doped 10^{16} cm^{-3} with a thickness of 2 μm and a carrier lifetime of 1 μs.

For the emitter Eq. (3.12a) can be used. However, the value of p_{n0} must be calculated using the mass action law with the effective doping level due to band-gap narrowing as in Eq. (4.17). In other words, $p_{neo} = n_i^2/10^{18} = 200$ cm^{-3}. The corresponding current from Eq. (3.12a) is:

$$I_{pE} = 1.6 \times 10^{-19} \times (10^{-4})^2 \times 1.25 \times 200/1 \times 10^{-4} = 4 \times 10^{-21} \text{A}$$

For the collector, Eq. (3.26) gives
$$I_{pc} = 1.6 \times 10^{-19} \times 10^{-8} \times [(1.4 \times 10^{10})^2/10^{16}] \times 2 \times 10^{-4}/10 \times 10^{-6} = 6.4 \times 10^{-22} \text{ A}.$$

These are the currents that would be measured by extrapolating the I–V characteristics from the forward-bias regions of each junction with the alternate junction at zero (or reverse) bias.

Actual reverse leakage currents: The actual reverse leakage current due to space-charge generation in the c–b junction, using Eq. (3.39) and assuming a depletion layer thickness equal to the 2 μm collector thickess will be:

$$I_{\text{scrgen}} = 1.6 \times 10^{-19} \times 10^{-8} \times 1.4 \times 10^{10} \times 1 \times 10^{-4}/1 \times 10^{-6} = 1.3 \times 10^{-15} \text{ or } 1.3 \text{ fA}$$

For the emitter–base junction the depletion layer will be thinner (the base is doped ten times more than the collector, and the maximum reverse bias voltage will be smaller). However, the lifetime in the emitter–base junction is in general less, and so the corresponding current can therefore be expected to be comparable in magnitude to the above value of 1.3 fA.

4.4.2 Current–voltage characteristics

Common-base characteristics

Figure 4.13 shows the ideal I_C versus V_{CB} characteristics of the BJT. Note that I_E is the parameter that is fixed for each curve. These are the common-base characteristics. The current gain in this configuration is defined as:

$$\alpha = I_C/I_E = \beta I_B/(I_B + I_C) \tag{4.31}$$

$$\alpha = \beta/(1 + \beta) \tag{4.32}$$

For a typical value of $\beta = 100$, we see that $\alpha = 0.99$.

Consider the V_{BE} values in Fig. 4.13 to be maintained as V_{CB} is reduced. The curves are almost horizontal, until V_{CB} changes sign and becomes comparable to V_{BE}. Then I_C decreases, as given by the Eq. (4.30), because the transistor is now in saturation. Eventually, as predicted by Eq. (4.30), I_C changes sign and the curves cross the x axis. The point P thus corresponds approximately to the curve labeled P in Fig. 4.11b (note that since some current exists due to the holes injected into the collector, $dn/dx = 0$ in Fig. 4.11b does not correspond exactly to $I_C = 0$).

Common-emitter characteristics

Figure 4.14 shows the ideal common-emitter characteristics. Since $V_{CB} = V_{CE} - V_{BE}$, the saturation region starts to the right of the y axis. Note that now the base current I_B is the

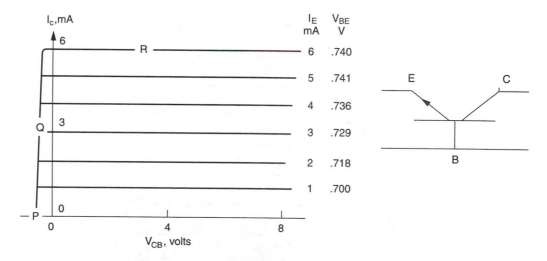

Figure 4.13 I_C versus V_{CB} common-base characteristics. Points labeled P, Q, R correspond to the points in Fig. 4.10(b). Note that constant values of I_E correspond to constant values of V_{BE} in the normal active regime. The common-base configuration is shown on right.

constant parameter for each curve. In reality, there is a finite slope dI_C/dV_{CE} due to the Early effect, to be discussed in Sections 4.5 and 4.6.

The curves in Fig. 4.13 can be deduced from those of Fig. 4.12 using the fact that $I_E = I_C + I_B$ and $V_{CB} = V_{CE} - V_{BE}$.

4.5 HIGH- AND LOW-CURRENT EFFECTS AND BREAKDOWN VOLTAGE

There are several effects that occur in the BJT at high currents and that lead to reduced current gain (and, as we shall see, reduced unity current gain frequency f_t). In general the device and/or circuit engineer needs to know the maximum usable collector current, beyond which the performance is degraded. The effects are as follows:

Base resistance voltage drop— emitter current crowding

When the voltage drop across the base resistance r_{bb} exceeds $V_t = kT/q$, the collector current density $J_n(y)$ becomes nonuniform, as shown in Fig. 4.15. Since $J_n(y) = J_{cs} \exp[V_{BE}(y)/V_t]$ and since $V_{BE}(y) = V_{BE}(0) - I_B r_{bb}$, it is clear that the collector current density will be higher at the edge of the base under the emitter than at the center. When the ratio $J_{n(edge)}/J_{n(center)}$ is significantly greater than unity, the transistor area is not being used effectively. This effect becomes serious when the $I_B r_{bb}$ voltage drop equals or exceeds kT/q (e.g., if this voltage drop equals 0.025 V, the current density ratio $J_{n(edge)}/J_{n(center)} = 2.7$). We can thus define a critical base current for the onset of emitter current crowding as:

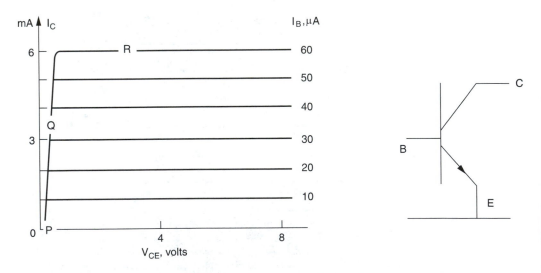

Figure 4.14 Common-emitter I_C versus V_{CE} characteristics. Points labeled P, Q, R corresspond approximately to the conditions labeled in Fig. 4.10(b). The common-emitter configuration is shown on the right.

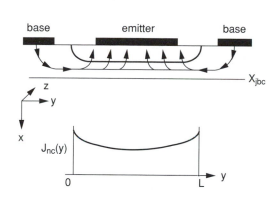

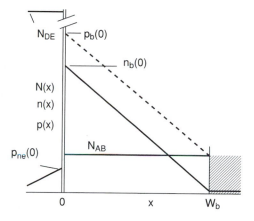

Figure 4.15 Emitter current crowding. The top diagram shows base current direction (arrows) as in Fig. 4.9. The bottom diagram represents the variation of vertical collector current density J_{nc} with horizontal distance y (after [7]). (See also CD-ROM BIPGRAPH example TRAN3B; "Vertical $J_c(y)$ vs y.")

Figure 4.16 High-level injection in the base region of the BJT. The base–collector space-charge layer is shown shaded, with an electron concentration $n_c \approx 0$. The emitter remains in low-level injection. Note the break in the y axis and the very narrow emitter–base space-charge layer.

$$I_{Bcrowd} = V_t / r_{bb} \tag{4.33}$$

Note that the transistor is still usable at higher currents, but this effect accentuates the other high-current effects discussed below.

High-level injection in the base

When the injected electron concentration $n_b(0)$ exceeds the background acceptor doping level N_{AB}, the base enters high-level injection, as shown in Fig. 4.16. This is similar to the narrow-base diode in high-level injection, as discussed in Chapter 3. The $I_C(V_{BE})$ law changes from an $\exp(V_{BE}/V_t)$ to an $\exp(V_{BE}/2V_t)$ dependence. Since the emitter is generally more heavily doped than the base, it does not enter high-level injection; so the base current (injected into the emitter) continues to increase as $\exp(V_{BE}/V_t)$. Thus the major consequence of this effect is that the current gain $\beta = I_C/I_B \propto \exp(-V_{BE}/2V_t)$ and falls off as $1/I_C$.

Kirk effect—base stretching

When the electron concentration n_c in the base–collector depletion layer exceeds the background doping level N_{epi}, the Poisson law solution is drastically altered:

$$\frac{dE}{dx} = (q/\epsilon)(N_{epi} - n_c) \tag{4.34}$$

and dE/dx thus changes sign. The corresponding critical collector current is given by:

$$I_K = q A_E v_s N_{epi} \tag{4.35}$$

where A_E is the emitter area $(L x B)$, v_s is the saturated drift velocity of electrons, and N_{epi} is the collector doping level (chosen for the required breakdown voltage).

In Fig. 4.17(a), $I_C \ll I_K (n_c \ll N_{epi})$, the base–collector depletion layer starts close to the base–collector junction, and the electric field peaks at this junction.

When the collector current I_C exceeds I_K, the electric field distribution peaks at the N_{epi}–N^+ interface, as shown in Fig. 4.17(b). Note that for constant V_{CB} (or constant V_{CE}), the area under the field versus depth curve remains constant. The field reduces rapidly to zero in the N^+ substrate or buried layer (according to Poisson's law, dE/dx is very large if N_D is very high). For $n_c = 2N_{epi}$, dE/dx has the same magnitude as case (a), but the sign is reversed in the N_{epi} region. The apparent junction is now shifted to the N_{epi}–N^+ junction. For the same V_{CB} the area under of $E(x)$ versus x is the same, and hence the field becomes zero at some point a distance W_K to the right of the base–collector junction. The neutral base (gradient dn/dx) now extends over the entire distance $W_b + W_K$, where W_b is the original (constant) value of the neutral base thickness.

Since $\beta \propto 1/(W_b + W_K)$, the current gain decreases. Also, as we shall see, the unity current gain frequency is such that $1/2\pi f_t$ is proportional to $t_{bb} = (W_b + W_K)^2/2D_n$. Therefore, f_t decreases even more rapidly. This effect starts when $n_c = N_{epi}$, that is, at a collector current $I_K = q A_E N_{epi} v_s$. This is the single most important high-current limit in the BJT.

Example of Kirk current calculation: Consider a $1 \times 1\ \mu m^2$ emitter BJT with a collector doped 10^{16} cm^{-3}.

Equation (4.35) gives

$$I_K = q A_e v_s N_{epi}$$
$$= 1.6 \times 10^{-19} \times (10^{-4})^2 \times 10^7 \times 10^{16} = 1.6 \times 10^{-4} = 0.16\,\text{mA}$$

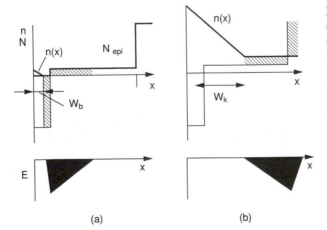

(a)   (b)

Figure 4.17 Illustration of Kirk effect (vertical base stretching) (a) $I_C \ll I_K$; (b) $I_C = 2I_K$. The top diagrams show the electron concenration $n(x)$ and the doping profile $N(x)$ with the base–collector depletion layer shaded. The lower diagrams show the electric field. (See also CD-ROM BIPGRAPH examples TR-NJ11, TR-NJ13, TR-NJ15, TR-NJ17; "$n(x)$ over all x" and "Electric field for all x.")

It may be noted that in some modern high-speed BJT the collector is lightly doped over a thickness W_{epi}, which is small enough for the depletion layer to *reach through* to the N^+ region even at low V_{CB} bias. In this structure the Kirk effect is determined by W_{epi}, and the value of J_K can be shown [7] to be equal to $2\epsilon V_{CB} v_s / W_{epi}^2$.

Quasi-saturation

The collector epitaxial layer doped at a value N_{epi}, thickness W_{epi} [see Fig. 4.18(a)], has a finite resistance to majority carrier electrons. This resistance may be written simply as:

$$R_{epi} = \rho_{epi} W_{epi} / A_E \tag{4.36}$$

where $\rho_{epi} = 1/(q \mu_n N_{epi})$.

When the voltage drop $V_{IR} = I_C R_{epi}$ due to I_C flowing in the epitaxial layer outside the depletion region becomes equal to the applied reverse V_{CB} value, the junction becomes forward biased. V_{CE} may appear to create a reverse-biased B–C junction, but the internal $V_{C'E}$ will be lower by an amount V_{IR}.

Example of quasi-saturation: Consider an epitaxial layer doped 10^{16} cm^{-3} and 1 μm thick outside the depletion layer, with an emitter area 1×1 μm^2.

The resistivity is $1/(1.6 \times 10^{-19} \times 1000 \times 1 \times 10^{16}) = 0.62$ ohm cm. The resistance R_{epi} from Eq. (4.36) is $0.62 \times 1 \times 10^{-4}/10^{-8} = 6200$ ohm. If $V_{CB} = 1$ V (reverse), it is clear from Fig. 4.18(b) that if I_C is increased above 0.16 mA, the junction will become forward biased (negative $V_{C'B}$). The transistor will be in saturation, even though it appears from the applied terminal bias to be in the normal active region. Again, β decreases when the forward V_{BC} bias becomes comparable to V_{BE}.

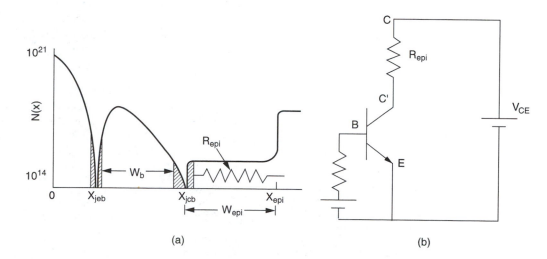

Figure 4.18 (a) Impurity profile recalling the epitaxial layer structure with a resistive region, giving rise to resistance R_{epi}. (b) Common-emitter configuration with the collector resistance R_{epi}.

The net result of the above effects is that the current gain falls off at high collector currents. We will see that the unity current gain frequency f_t is also directly related to base width W_b and to excess minority carrier base charge and so degrades simultaneously with β.

Low-current gain fall-off

In Chapter 3 we showed that at low forward bias the current in a diode was due primarily to space-charge recombination—the free carriers in the forward-biased space-charge layer recombining with a lifetime τ. We saw that the corresponding current was proportional to $\exp(V_a/mV_t)$ where m was close to 2. In the case of the transistor, the same current exists and becomes important at low forward V_{BE} bias. It is most important to note, however, that its effect is only to increase the base current I_B. It is the base that supplies both the current injected into the emitter and the current due to recombination in the e–b space-charge layer. Thus at low forward bias we have:

$$I_B \propto \exp(V_{BE}/mV_t) \tag{4.37}$$

The corresponding current gain is:

$$\beta = I_C/I_B \propto \exp(V_{BE}/V_t)/\exp(V_{BE}/mV_t) \tag{4.38}$$

$$\propto \exp[V_{BE}(1 - 1/m)/V_t] \tag{4.39}$$

For $m = 2$, a typical value, this gives a current gain proportional to the square root of collector current. The general shape of the β versus I_C curve is thus as shown in Fig. 4.19. As the current increases, the gain becomes limited by the constant value already derived for injection of holes into the emitter. At very high currents, one of the above-discussed effects (high-level injection, Kirk effect, quasisaturation) causes the gain to fall off.

A frequently used method of presenting all the important current-dependent information of the BJT is the Gummel plot, shown in Fig. 4.20. This shows both I_C and I_B on a logarithmic scale, plotted versus V_{BE} on a linear scale. The separation between the I_C and I_B curves is the current gain. Notice how the base current plot has a decreased gradient at low V_{BE} due to e–b space-charge recombination. The change in slope of the I_C plot at high currents is due to the high-current phenomena discussed above—mainly vertical base stretching (Kirk effect) plus the $I_B r_{bb}$ voltage drop.

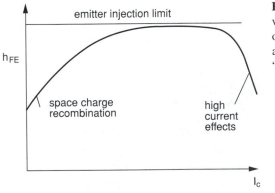

Figure 4.19 Plot of current gain $h_{FE}(\beta)$ versus I_C; the emitter injection component of β is shown by the horizontal line. (See also CD-ROM BIPGRAPH example TRAN1; "BETA DC vs I_c.")

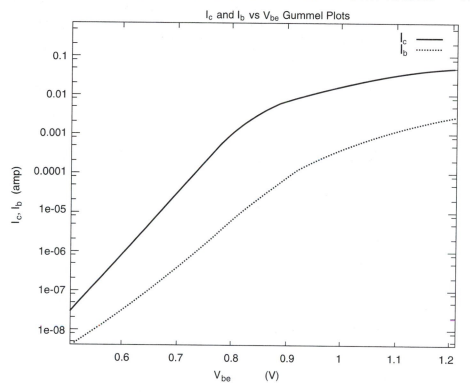

Figure 4.20 Gummel plot of I_C and I_B (log scales) versus V_{BE} (linear scale) (*generated by the BIPOLE3 computer simulation program,* [8]). (See also CD-ROM BIPGRAPH example TRAN1; "Collector and base currents vs V_{be}.")

Avalanche breakdown voltage

The breakdown voltage BV_{CBO} for the base–collector junction is identical to that of the P–N diode analyzed in Chapter 3. We can thus take the result directly and write:

$$BV_{CBO} = \epsilon E_{br}^2/(2qN_{epi}) \qquad (4.40)$$

This is the starting point for designing a BJT—the collector doping level N_{epi} and thickness W_{epi} are chosen to give the required value of breakdown voltage. There is, however, a second breakdown limitation in the BJT, when operating in the common-emitter configuration, in which case the maximum value of V_{CE}, called BV_{CEO}, is restricted to a considerably smaller value than BV_{CBO}. The following treatment is simplified and nonrigorous, but will give the reader an idea of this additional voltage limitation.

The current passing through the base collector space-charge layer is multiplied by the avalanche multiplication factor M given by Eq. (3.40):

$$M = 1/[1 - (V_{CB}/BV_{CBO})^n] \tag{4.41}$$

The collector current is now given in terms of the emitter current I_E by:

$$I_C = -M\alpha I_E + I_s' \tag{4.42}$$

where α is the common-base current gain and I_s' is related to the reverse leakage current of the base–collector junction.

For the case of an open-circuit base contact, where $I_B = 0$, we have $I_C = -I_E$, and the collector current becomes:

$$I_C = I_s'/(1 - M\alpha) \tag{4.43}$$

Since α is typically about 0.99, it is clear than at V_{CB} values well below BV_{CBO}, M can reach a value of 1.01, and the collector current is unlimited. The actual maximum voltage limit can be determined by setting $M\alpha = 1$; thus:

$$\alpha/[1 - (V_{CB}/BV_{CBO})^n] = 1 \tag{4.44}$$

setting $V_{CB} = BV_{CEO}$ gives:

$$BV_{CEO} = BV_{CBO}(1 - \alpha)^{1/n} \tag{4.45}$$

$$BV_{CEO} = BV_{CBO}/\beta^{1/n} \tag{4.46}$$

For typical values of $\beta = 100$, $n = 2$, we see that BV_{CEO} can be of order one-tenth BV_{CBO}, although in practice BV_{CBO} is reduced by breakdown in the diffused sidewall regions of the junction; so this factor is also reduced, typically to about $\frac{1}{3}$. The engineering design of a BJT starts with the choice of collector doping level (4.40) and corresponding thickness (2.62) for the required breakdown voltages BV_{CBO}, BV_{CEO}.

Dependence of collector current on collector voltage (the Early effect)

Figure 4.21 shows the effect on the base width W_b on collector voltage. As V_{CB} (or V_{CE}) increases, in the reverse-bias direction, the c–b depletion layer extends more into the base region, as shown in Fig 4.21, thus decreasing the value of neutral base width W_b. The gradient dn/dx is thus increased in magnitude, and for constant V_{BE}, the collector current I_C thus increases, producing a finite dI_C/dV_{CE} gradient as shown in the complete I_C–V_{CE} characteristics of Fig. 4.22. The set of curves extrapolates backwards to intersect the x axis at approximately the same point, and the reverse intercept is called the *Early voltage*, referred to as V_A or V_{AF} in SPICE CAD models. The double-valued characteristics in the region of BV_{CEO} are due to avalanche multiplication, combined with thermal feedback effects.

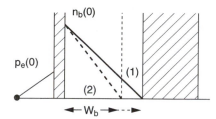

Figure 4.21 Explanation of finite output conductance—the Early effect. Case (1) is for low V_{CE}; case (2) is for a higher V_{CE}.

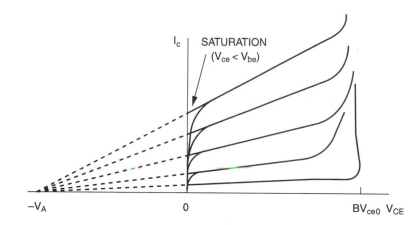

Figure 4.22 I_C versus V_{CE} characteristics showing finite output conductance and Early voltage definition.

4.6 BJT SMALL-SIGNAL PROPERTIES

4.6.1 Small-signal equivalent circuit

We are now in a position to derive the small-signal equivalent circuit for the BJT. This is shown in Fig. 4.23 and is normally referred to as the hybrid π model (π because of the "shape" of the topology and "hybrid" because the several parameters have different "dimensions" as opposed to y or z parameter models, used in the early days of BJT circuit analysis). It is particularly useful for common-emitter circuit calculations.

Low-frequency components

From the above discussion on base resistance, it is clear that we should include this resistance r_{bb} in series with the base terminal. The small-signal transconductance g_m (defined in terms of the small-signal output current i_c and input voltage v'_{be} as i_c/v'_{be}, or in terms of incremental dc quantities by dI_C/dV'_{BE}) of the BJT may be obtained directly from our previous analysis for collector current:

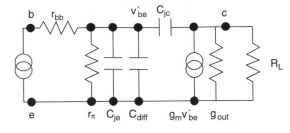

Figure 4.23 Hybrid π small-signal equivalent circuit of the BJT. A current source is shown at the input and a load resistance at the output.

$$I_C = I_{cs} \exp(V'_{BE}/V_t) \qquad (4.47)$$

$$g_m = \frac{dI_C}{dV'_{BE}} = (I_{cs}/V_t) \exp(V'_{BE}/V_t) \qquad (4.48)$$

Substituting to give the result in terms of collector current gives:

$$g_m = I_C/V_t \qquad (4.49)$$

This is a very important property of the BJT. It tells us that the transconductance may be made very large simply by increasing the forward V_{BE} bias. It is the main reason why the BJT is able to drive large capacitive loads and yet maintain its high speed in digital circuits. The upper usable limit on collector current is determined by the onset of the various effects discussed in Section 4.5.

Example: For a collector current $I_C = 1$ mA at 17°C, the transconductance $g_m = 10^{-3}/0.025 = 0.04 = 40$ mA/V.

The small-signal output conductance is the slope of the I_C versus V_{CE} curve of Fig. 4.22.

$$g_{out} = \frac{dI_C}{dV_{CE}} \qquad (4.50)$$

Using the definition of Early voltage enables this to be written as:

$$g_{out} = I_C/(V_{CE} + V_A) \qquad (4.51)$$

There is a simple but important relationship between the Early voltage and the base charge or Gummel integral $G_b = N_{AB} W_b$. From Eq. (4.30) in the normal active regime (V_{BC} negative) we can write:

$$I_C = K/G_b \qquad (4.52)$$

where K is a constant. Equation (4.50) can be rearranged as:

$$g_{out} = \frac{dI_C}{dG_b} \frac{dG_b}{dV_{CE}} \tag{4.53}$$

We note that the base–collector depletion layer capacitance per unit area $C'_{jc} = dQ_B/dV_{CB}$ may be written as $q\,dG_b/dV_{CB} = q\,dG_b/dV_{CE}$ (for constant V_{BE}). Using this fact, differentiating Eq. (4.52) and equating the result to Eq. (4.51) gives:

$$V_A + V_{CE} = qG_b/C'_{jc} \tag{4.54}$$

Example of output conductance and Early voltage: For the BJT used in previous sections, $qN_{AB}W_b = 1.6 \times 10^{-19} \times 10^{17} \times 10^{-4} = 1.6 \times 10^{-6}$.

The value of $C_{jc'}$ can be calculated at a depletion-layer thickness $w_{scl} = 1\,\mu m$; $C_{jc'} = \epsilon/w_{scl} = 10^{-12}/10^{-4} = 10^{-8}\,F/cm^2$.

Equation (4.54) thus gives $V_A + V_{CE} = 1.6 \times 10^{-6}/10^{-8} = 160$ V, a very typical value for medium-performance BJTs.

The output conductance at $I_C = 0.1$ mA, $V_{CE} = 5$ V, is $I_C/(V_A + V_{CE}) = 0.1 \times 10^{-3}/(160 + 5) = 6 \times 10^{-7}$. In other words, the small-signal output resistance $r_{out} = 1/g_{out} = 1.6$ Mohm.

Since the base charge qG_b is reduced as V_{CB} is increased, it follows that the Early voltage is not rigorously constant. Nevertheless, the approximation is good for many situations and is a useful SPICE modeling concept.

The small-signal resistance r_π in Fig. 4.23 defined by the reciprocal of the slope of the I_B–V_{BE} characteristic of the base–emitter "diode" law is defined by:

$$r_\pi^{-1} = \frac{dI_B}{dV'_{BE}} \tag{4.55a}$$

$$= \frac{dI_B}{dI_C} \frac{dI_C}{dV'_{BE}} \tag{4.55b}$$

$$= g_m/h_{fe} \tag{4.56}$$

or

$$r_\pi = h_{fe}/g_m \tag{4.57}$$

where we have introduced the ac current gain $h_{fe} = dI_C/dI_B$. Based on the previous analysis, h_{fe} (lower-case subscripts to distinguish it from the dc current gain h_{FE}) is the same as the dc current gain h_{FE} or β. This is approximately true, but the two quantities are not quite the same because of the nonlinear dependence of current gain on collector current outlined in Section 4.5.

Example of r_π calculation: For a collector current $I_C = 1$ mA at 17°C, and $h_{fe} = 100$, $r_\pi = 100/(1 \times 10^{-3}/0.025) = 2.5$ kohm.

High-frequency components

Because of the two junctions, each with its space-charge layer, there will be two depletion layer capacitances as shown in Fig. 4.22. The emitter–base depletion-layer capacitance C_{je} is between the emitter and base terminals; the collector–base depletion-layer capacitance C_{jc} is between collector and base terminals. These capacitances obey the voltage dependence discussed in Chapter 3 for the diode.

$$C_{je} = C_{jeo}/(1 - V_{BE}/V_{bie})^{MJE} \tag{4.58}$$

$$C_{jc} = C_{jco}/(1 - V_{BC}/V_{bic})^{MJC} \tag{4.59}$$

where C_{jco}, C_{jeo} are the zero-bias values of the capacitances; V_{bie}, V_{bic} are the built-in barrier potentials of each junction. MJE $\approx \frac{1}{3}$ for the e–b junction, which under forward bias approximates a linear graded junction as discussed in Section 2.4.6 because of the impurity profile variation with depth seen in Fig. 4.3. MJC $\approx \frac{1}{2}$, since under reverse bias the collector doping is constant and the junction approximates a P^+N asymmetrical junction.

The diffusion capacitance is similar to that of the narrow-base diode discussed in Chapter 3.

$$C_{diff} = \frac{dQ_{nb}}{dV_{BE}} \tag{4.60}$$

where Q_{nb} is the charge of the electron distribution in the base (cf. Fig. 4.6). Rearranging terms gives:

$$C_{diff} = \frac{dQ_{nb}}{dI_C}\frac{dI_C}{dV_{BE}} \tag{4.61}$$

$$= (Q_{nb}/I_C)g_m \tag{4.62}$$

where we neglect the voltage drop across the base resistance r_{bb} in defining g_m (this is not an approximation; it simply means that r_{bb} must be included in the complete circuit analysis). As in the narrow-base diode of Chapter 3, the electron charge is proportional to the electron current. The ratio is obtained by substituting for the charge $Q_{nb} = \frac{1}{2}qA_En_b(0)W_b$ and for the current $I_C = qA_ED_nn_b(0)/W_b$ to obtain:

$$C_{diff} = g_m(W_b^2/2D_n) \tag{4.63}$$

$$= g_mt_{bb} \tag{4.64}$$

where

$$t_{bb} = W_b^2/2D_n \tag{4.65}$$

is the base transit time, that is, the time it takes for the electrons to diffuse across the base region width W_b. This neglects the free-electron charge Q_c in the c–b space-charge layer.

From Fig. 4.6 we can deduce that this charge is equal to $q A_E n_c w_{scl}$, where w_{scl} is the depletion-layer thickness. A first-order analysis assuming that the electrons travel at their saturated drift velocity v_s, would indicate a corresponding delay time $t_{scl} = Q_c/I_C = w_{scl}/v_s$. In fact, a rigorous analysis [7] gives:

$$t_{scl} = w_{scl}/2v_s \tag{4.66}$$

The total forward delay time is usually referred to as t_f, which is the sum of t_{bb}, t_{scl}, and the delay t_e due to minority carrier charge injected into the emitter. In this case we can write:

$$C_{diff} = g_m t_f \tag{4.67}$$

where neglecting t_e

$$t_f = t_{bb} + t_{scl}. \tag{4.68}$$

4.6.2 Frequency response and unity current gain frequency f_t

From Fig. 4.22 we may write the frequency dependence of the small-signal output current versus input current. For the special case of a *short-circuit output* ($R_L = 0$) this is denoted by $h_{fe}(\omega)$. The short-circuit output condition is used because it yields a parameter that is independent of the circuit and characterizes only the active BJT device. It also corresponds with the original definition of forward current gain using the h parameter notation (see below). Denoting the voltage at the input node by v'_{be} as shown in Fig. 4.23, we have:

$$i_c = g_m v'_{be} \tag{4.69}$$

The input current is given by:

$$i_b = v'_{be}/Z_\pi \tag{4.70}$$

where Z_π is the parallel combination of r_π, C_{diff}, C_{je}, and C_{jc}. Note the presence of C_{jc} due to the fact that we are considering a short-circuit output; so the right-hand side of C_{jc} is grounded and thus appears in parallel with C_{je} and C_{diff}. The current gain is thus:

$$h_{fe}(\omega) = i_c/i_b = g_m Z_\pi \tag{4.71}$$

$$= g_m r_\pi/(1 + j\omega C_{\pi t} r_\pi) \tag{4.72}$$

where we define $C_{\pi t} = C_{je} + C_{jc} + C_{diff}$. Substituting from the previously derived expression for $r_\pi = g_m/h_{fe}$ gives:

$$h_{fe}(\omega) = h_{fe}/(1 + j\omega C_{\pi t} r_\pi) \tag{4.73}$$

This is plotted in Fig. 4.24. The -3 dB frequency, when the amplitude is reduced by $1/\sqrt{2}$, is called the beta cut-off frequency and is given by:

$$f_{c\beta} = 1/(2\pi C_{\pi t} r_\pi) \tag{4.74}$$

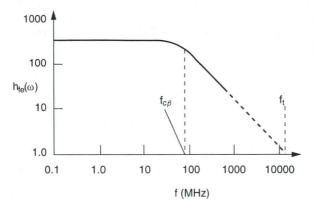

Figure 4.24 Small-signal current gain $h_{fe}(\omega)$ versus frequency using a log–log plot.

Beyond this frequency the gain falls off at -6dB per octave (-20 dB per decade). The value at which the extrapolated gain becomes unity is referred to as the *unity current gain frequency*, the *transition frequency*, or simply as the f_t of the transistor. Note that the extrapolation must be performed from frequencies considerably less than f_t since the hybrid model fails before f_t is reached and the constant slope is no longer observed. From the above result, assuming (as is always the case for a good transistor) that the low-frequency value of h_{fe} is much greater than unity, setting $h_{fe}(\omega) = 1$ in Eq. (4.73), we obtain:

$$f_t = g_m/(2\pi C_{\pi t}) \tag{4.75}$$

$$= h_{fe} f_{c\beta} \tag{4.76}$$

It is convenient to examine the current dependence of f_t by taking its reciprocal:

$$1/2\pi f_t = (1/g_m)(C_{je} + C_{jc}) + C_{diff}/g_m$$

This may be written as:

$$1/2\pi f_t = (V_t/I_C)(C_{je} + C_{jc}) + t_f \tag{4.77}$$

Since the depletion-layer capacitances vary only slowly with dc current, they may be considered as approximately constant, if evaluated close to normal operating bias. The plot of f_t versus I_C is then as shown in Fig. 4.25. The high-current fall-off is due to the Kirk effect or quasi-saturation. Note that high-level injection does not degrade the f_t. Note that the peak value of f_t approaches the value limited by the forward delay time t_f (of which the dominant components are $t_{bb} + t_{scl}$).

Common-base configuration

In the common-base configuration (Section 4.4.2, Fig. 4.13) the input resistance is $dV_{BE}/dI_E = (dV_{BE}/dI_B)(dI_B/dI_C)(dI_C/dI_E) = (r_\pi/h_{fe})\alpha$. This is is approximately r_π/h_{fe}. Since the input capacitance is still between base and emitter terminals, it is equal to $C_{diff} + C_{je}$. It

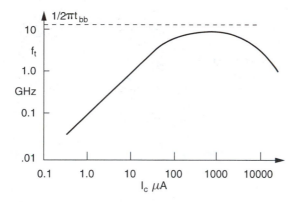

Figure 4.25 f_t versus I_C for a BJT. (See also CD-ROM BIPGRAPH example TRAN1; "f_t vs I_c.")

is therefore evident that the 3 dB *common-base current gain cutoff frequency*, called f_α, is given by:

$$f_\alpha \approx h_{fe} f_{c\beta} = f_t \tag{4.78}$$

For this reason, the common-base configuration is sometimes used in microwave applications (where the low input impedance is not such a severe problem).

Maximum oscillation frequency

There is a second very important high-frequency figure of merit, called the maximum oscillation frequency f_{maxosc} or simply f_{max}. This is the frequency at which the small-signal power gain (calculated from the above hybrid π model) becomes unity. Appendix L gives the derivation based on circuit analysis, the result of which is:

$$f_{maxosc} = [f_t/(8\pi r_{bb} C_{jc})]^{1/2} \tag{4.79}$$

This figure of merit has the advantage of including two important device quantities, the base resistance and the collector junction capacitance, both of which seriously affect many aspects of circuit performance. Thus a good BJT will have not only a high f_t but also a high f_{maxosc}. This means that the base width cannot be decreased to a very small value, because this would lead automatically to a very high value of r_{bb}, as can be seen from Eq. (4.23).

Example of f_{max} calculation: Consider the values used in the previous examples in this chapter with in addition a collector junction area equal to 4 times the emitter area.

Taking the hole mobility from Fig. 2.11 at a doping level 10^{17} cm^{-3} gives approximately 300 cm^2/V s. Using this value in the base sheet resistance equation (4.24) gives $R_{be/sq} = 2.1$ kohm/sq.

The base resistance is given by [Eq. (4.24)] as: $r_{bb} = 2100(L/12B) = 2100/12 = 175$ ohm.

Using the same depletion layer as in the Early voltage example (1 μm) gives $C_{jc} = \epsilon A_c/w_{scl} = 10^{-12} \times 4 \times 10^{-8}/10^{-4} = 4 \times 10^{-16} = 0.4$ fF.

From Eq. (4.65) the base transit time $t_{bb} = (1 \times 10^{-4})^2/(2 \times 20) = 2.5 \times 10^{-10}$ s.
From Eq. (4.66) $t_{scl} = 10^{-4}/2 \times 10^7 = 0.5 \times 10^{-11}$ s.
We will assume that f_t is the peak value given by
$1/2\pi(t_{bb} + t_{scl}) = 1/[6.28(255 \times 10^{-12}) = 624$ MHz.
The peak value of maximum oscillation frequency (4.79) is thus:

$$f_{maxosc} = [624 \times 10^6/(8\pi \times 175 \times 0.4 \times 10^{-15})]^{1/2} = 18.6 \text{ GHz}$$

4.6.3 ECL propagation delay time

The main logic circuit for bipolar transistors is now emitter-coupled logic (ECL). The basic relationships for this important logic family are given in Appendix M. The result of main interest is for an optimized circuit in which the delay is minimized. In this case the minimum delay is given with some approximations by:

$$t_{pd,min} = r_{bb}(C_{je} + 3C_{jc}) + \sqrt{3}/\omega_{max} + t_f \tag{4.80}$$

Note that for a BJT operating near the peak of the f_t versus I_C curve, $f_t \approx 1/2\pi t_f$ and thus we have:

$$t_{pd,min} \approx R_b C_{je} + 0.27/f_{max} + 1/2\pi f_t \tag{4.81}$$

It is not therefore surprising that a transistor designed and biased for maximum f_{max} gives near-minimum ECL propagation delay time. Figure 4.26 shows typical plots of f_{maxosc} and t_{pdECL}, where it is seen that as f_{maxosc} rises with I_C, the delay time falls.

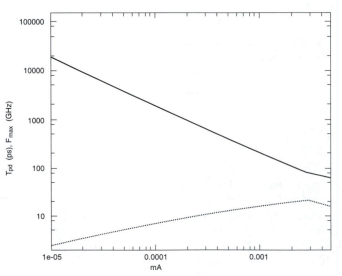

Figure 4.26 Plots of ECL propagation delay time and maximum oscillation frequency versus collector current I_C for a transistor with $f_t = 25$ GHz (top curve is t_{pd}; lower curve is f_{max}) generated using BIPOLE3 [8]. (See also CD-ROM BIPGRAPH example TRAN1; "ECL delay and F$_{max}$ vs I_c.")

4.6.4 The small-signal *h* parameters

The network *h* parameters are a black box representation of any active two-port device. Other two-port parameters are the *z*, *y*, *s* parameters. For the system shown in Fig. 4.27 for the common-emitter BJT configuration, the following definitions apply:

$$v_{be} = h_{ie}i_b + h_{re}v_{ce} \tag{4.82}$$

$$i_c = h_{fe}i_b + h_{0e}v_{ce} \tag{4.83}$$

From this pair of equations, the definitions of the two important *h* parameters follow:

$$h_{ie} = (v_{be}/i_b)|_{vce=0}; \qquad h_{fe} = (i_c/i_b)|_{vce=0} \tag{4.84}$$

In other words, h_{ie} is the input impedance ($r_{bb} + Z_\pi$) with a short-circuit output; h_{fe} is the current gain with a short-circuit output (as already discussed). Note that by "short-circuit output" we mean short circuit for ac signals only ($v_{ce} = 0$). This can be achieved by placing a large capacitance at the output node. The dc bias must, of course, be applied to the transistor. Since in the BJT, the output is a high impedance, the output "short" condition is not really too important. The current gain and the input resistance remain practically independent of the ac output voltage. However, at high frequencies, it is important to take the above definitions into account (as, for example, in the f_t derivation, where we saw that C_{jc} is in parallel with C_{je} and C_{diff}). The remaining two *h* parameters are defined by:

$$h_{re} = (v_{be}/v_{ce})|_{ib=0}; \qquad h_{0e} = (i_c/v_{ce})|_{ib=0} \tag{4.85}$$

In other words, h_{re} is the inverse voltage gain, and h_{0e} is the output admittance, both defined with an open-circuit input ($i_b = 0$). At low frequencies there is only a second-order effect, which makes both these terms non-zero, but for most practical design situations they can be set to zero. The non-zero values are related to the Early effect discussed in Section 4.4, and h_{0e} is in fact the gradient dI_C/dV_{CE} of the dc $I_C - V_{CE}$ characteristics.

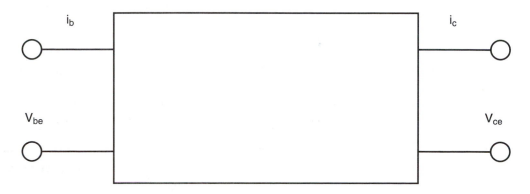

Figure 4.27 Blackbox representation of BJT in a common-emitter configuration.

4.6.5 Summary of BJT properties and detailed example

Table 4.1 summarizes the main circuit properties of the BJT. Typical values are given for a transistor operating at $I_C = 1$ mA, with $h_{fe} = 100$ and $f_t = 1$ GHz.

TABLE 4.1
Summary of BJT circuit properties for common emitter (CE) and common base (CB)

	r_{in}		r_{out}		A_i		A_v		$f_{3\,dB}$	
CE	$r_\pi = h_{fe}V_t/I_C$	2.5 K	V_A/I_C	100 kohm	h_{fe}	100	$g_m R_L$	100	f_t/h_{fe}	10 MHz
CB	V_t/I_E	25 ohm	$h_{fe}V_A/I_C$	10 Mohm	$h_{fe}/(1+h_{fe})$	0.99	$g_m R_L$	100	f_t	1 GHz

Example of VLSI BJT

The following details are typical of a VLSI BJT:

Emitter dimensions 3×3 μm^2; $X_{jeb} = 0.1$ μm; $X_{jbc} = 0.2$ μm; effective emitter doping level (including BGN effects) 10^{18} cm^{-3}; average base doping level 10^{17} cm^{-3}; collector doping level $N_{epi} = 10^{16}$ cm^{-3}, $W_{epi} = 1.0$ μm; $D_{n(base)} = 20$ cm^2/s; $D_{p(base)} = 10$ cm^2/s; $D_{p(emitter)} = 1$ cm^2/V s. Assume operation at $V_{CB} = 2.0$ V.

If we neglect the depletion layer regions in the base, $W_b = X_{jbc} - X_{jeb} = 0.1$ μm. We find the following results:

From Eq. (4.17), the current gain is:

$$\beta = (D_n N_{Deff} W_e)/(D_p N_{AB} W_b) = 200$$

The base transit time, Eq. (4.65) is $t_{bb} = W_b^2/2D_n = 2.5 \times 10^{-12}$ s = 2.5 ps.

The base–collector depletion layer thickness from Eq. (3.46): $w_{scl} = [2\epsilon V_{jtot}/qN_{epi}]^{1/2}$ for $V_{jtot} = V_{CB} + V_{bi} \approx 2.7$ V is = 0.5 μm; the corresponding delay time, Eq. (4.66), $t_{scl} = w_{scl}/2\,v_s = 2.5$ ps.

The maximum possible f_t, at a current high enough for $(1/g_m)(C_{je} + C_{jc})$ in Eq. (4.77) to be negligible, is thus: $f_t = 1/[2\pi(t_{bb} + t_{scl})] = 32$ GHz. The corresponding beta cut-off frequency is $32/200 = 0.16$ GHz, or 160 MHz.

The base–collector junction capacitance is $C_{jc} = \epsilon A_c/w_{scl}$; the collector–base junction area is typically about 4 times the emitter area (because of space taken up by the base contacts).

Thus $C_{jc} = 7.2$ fF (1 fF = 10^{-15} F).

To calculate the value of emitter–base junction capacitance C_{je}, we can assume that near normal forward bias the total junction voltage V_{jtot} is of order 0.1 V, i.e., V_{BE} is forward biased to within 0.1 V of the built-in barrier potential. The e–b depletion layer thickness is thus estimated from Eq. (3.46) as:

$W_{sceb} = [2\epsilon V_{jtot}/(qN_{AB})]^{1/2} = 0.035$ μm and $C_{je} = \epsilon A_e/w_{sceb} = 25$ fF.

The value of f_t at some low current, say, 1 μA, is therefore: $f_t = (I_C/V_t)/(C_{je} + C_{jc}) = 1.2$ GHz.

These values are slightly optimistic (by a factor of order 2 times) because we have neglected various parasitics such as sidewall junction capacitance and emitter delay time, but the values are nevertheless indicative of trends in modern VLSI structures.

It is also possible to have the collector as the common node connected directly to the supply voltage V_{CC} with the output taken from the emitter with a resistance R_L between emitter and ground; this is the common collector mode of operation. The common collector is used mainly as a near unity voltage gain "buffer" with a high input resistance (of order $h_{fe}R_L$) and low output resistance (of order $1/g_m$).

4.7 THE INTEGRATED CIRCUIT BIPOLAR TRANSISTOR

4.7.1 The vertical *NPN* transistor

The integrated circuit (IC) *NPN* transistor differs from the discrete version only in the arrangement of the collector and the fact that for the IC BJT a lightly doped P^- substrate is used. In the IC device, the N^+ substrate is replaced by an N^+ "buried layer," with the collector contact brought out to the surface. The top mask diagram and cross section of this arrangement is shown in Fig. 4.28. The principal difference between this device and the discrete BJT lies in the extra series collector resistance in the N^+ buried layer (see Chapter 8 for technology details). Often only one base contact is used, to save in layout area (at the expense of a larger base resistance). The deep P^+ diffusion surrounding the transistor on the P^- substrate serves to isolate the collectors of individual transistors electrically (by two reverse-biased PN junctions back to back).

4.7.2 The lateral *PNP* transistor

It is often necessary in circuits to have available a *PNP* transistor on the same chip as an *NPN* transistor. This may be achieved very simply using the lateral *PNP* structure. It requires no

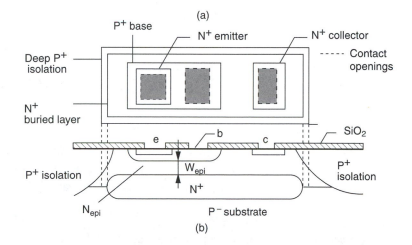

Figure 4.28 Integrated circuit BJT structure (a) mask layout; (b) cross section. Metal contacts are shown shaded. (Reprinted with permission from D. J. Roulston, *Bipolar Semiconductor Devices*, McGraw–Hill, New York, 1990.

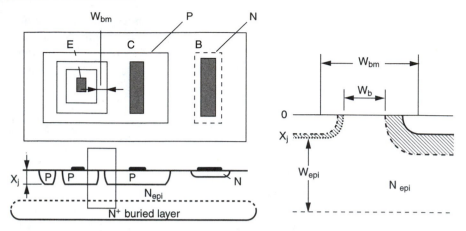

Figure 4.29 The lateral *PNP* transistor: (a) mask layout and cross section (metal contacts shown shaded); (b) enlarged cross section of active base region. Note that the *P* collector diffusion surrounds the *P* emitter diffusion.

aditional processing steps, and although its electrical performance is mediocre, it is widely used. Figure 4.29 shows the essential features.

In normal active region operation the emitter is forward biased and the collector reverse biased. Holes injected from the emitter sideways are collected at a horizontal distance W_b. The enlarged diagram shows that W_b is the distance between the two depletion layers (shown shaded). This is less than the lateral mask spacing W_{bm} due to: (1) sideways diffusion of the *P* layer, (2) sideways extension of the two depletion layers, particularly the reverse-biased collector–base depletion layer. The significant electrical parameters are the current gain and the f_t.

The collector current may be written as:

$$I_C = q D_p B_p X_j p_n(0)/W_b \tag{4.86}$$

where $B_p X_j$ is the vertical area around the emitter, perimeter B_p, $p_n(0)$ is the injected hole concentration, an exponential function of V_{BE} as in the *NPN* device.

Current gain

One component of base current is due mainly to the hole charge under the emitter in the epitaxial layer thickness W_{epi} recombining with a lifetime $\tau_{p.}$. This is similar to the N_{epi} layer of the P^+NN^+ diode of Chapter 3. This component of base current may be written as:

$$I_{B1} = Q_p/\tau_{p.} = q(B_p/4)^2 W_{epi} p_n(0)/\tau_{p.} \tag{4.87a}$$

A second component is due to recombination at the surface of the base over the distance W_b with a perimeter B_p. This current can be calculated by using the surface recombination velocity S_{0x} and the average hole concentration $p_{av} = p_n(0)/2$. Surface recombination velocity defines the rate at which minority carriers are "absorbed" into the surface and typically lies in the range 10 to 10^5 cm/s. The corresponding base current is thus using Eq. (2.79):

$$I_{B2} = q[p_n(0)/2]W_b B_p S_{0x} \qquad (4.87b)$$

The current gain is thus:

$$\beta = I_C/(I_{B1} + I_{B2}) = [16(X_j D_p \tau_p)/(W_b B_p W_{epi})]/[1 + 16\,(W_b/W_{epi})(S_{0x}\tau_p/2B_p)] \quad (4.88)$$

Substituting for the lifetime in terms of the hole diffusion length $L_p = \sqrt{D_p \tau_p}$ enables this to be written as:

$$\beta = (X_j/W_b)(L_p/B_p)(L_p/W_{epi})/[1/16 + (W_b/W_{epi})(S_{0x}\tau_p/2B_p)] \qquad (4.89)$$

Example of lateral *PNP* gain:

Junction depth $= 1.0\,\mu m$, $W_b = 1.0\,\mu m$, $W_{epi} = 2.0\,\mu m$, $L_p = 30\,\mu m$ ($\tau_p = 0.9\,\mu s$), $S_{0x} = 10^4$ cm/s. For a 5 μm square emitter diffusion, the perimeter $B_p = 20\,\mu m$. Substituting gives:

$$\beta = (1/1)(30/20)(30/2)/[1/16 + (1/2)(10^4 \times 0.9 \times 10^{-6}/2 \times 20 \times 10^{-4})$$
$$= 22.5/(0.0625 + 11.25) = 2.0$$

The gain is further reduced by injection into the emitter as in a vertical *NPN* transistor and by hole charge diffusing and recombining outside the area under the emitter, as shown in Fig. 4.12 for the upward-operating BJT (the P to N_{epi} junction is forward biased in both cases).

High-frequency cut-off

The f_t may be deduced using Eqs. (4.62)–(4.65) and (4.77) and defining a diffusion delay time as:

$$t_{diff} = Q_p/I_C \qquad (4.90)$$

Substituting for charge and current from the above equations gives:

$$t_{diff} = (B_p W_{epi} W_b)/(X_j D_p) \qquad (4.91)$$

To better see the magnitudes this can be rewritten as:

$$t_{diff} = (W_b^2/2D_p)2(B_p/X_j)(W_{epi}/W_b) \qquad (4.92)$$

The first term is identifiable as the diffusion transit time across the base as in Eq. (4.65). We see that even if W_b is kept small by careful choice of mask dimensions allowing for sideways diffusion, the factor $2(B_p/X_j)(W_{epi}/W_b)$ will, using the numbers in the above example, be large, of order 80. The lateral *PNP* transistor will therefore usually have a very large value of t_{diff} and hence a small f_t (peaking at a value $1/2\pi t_{diff}$).

Example of f_t calculation for lateral *PNP* transistor:
     Using the previous numbers and Eq. (4.92),
$t_{\text{diff}} = (10^{-4})^2/(2 \times 10) \times 2 \times (20/1)(2/1) = 2 \times 10^{-8}$ s.
     The maximum possible value of f_t is therefore $f_t = 1/2\pi t_{\text{diff}} = 8$ MHz.

4.8 SWITCHING BEHAVIOR

Figure 4.30 shows the BJT with a voltage source switching the base through a resistance R_s. The base–emitter diode behaves in a very similar manner to the diodes discussed in Chapter 3 during switching. The base–emitter voltage and current waveforms are shown in Fig. 4.31.

Turn-on delay

There is an initial delay due to the fact that at the start of the pulse the junction can be reverse biased. This delay can be calculated by conventional circuit techniques, assuming an average constant value of the C_{je} junction capacitance (chosen between its reverse-bias value and some small forward bias of about 0.5 V). Defining the time constant $t_{\text{rc}} = R_s C_{\text{jeav}}$, the initial turn-on delay time is given by:

$$t_{0n1} = t_{\text{rc}} \ln[(V_2 - V_1)/(V_2 - V_{0n})] \tag{4.93}$$

where V_{0n} may be taken as within about 0.1 V of the final V_{BE} value, that is, say, 0.6 V. For the common case where V_2 is large (to turn the transistor on rapidly), this simplifies to:

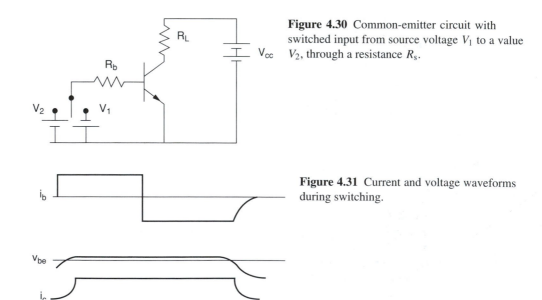

Figure 4.30 Common-emitter circuit with switched input from source voltage V_1 to a value V_2, through a resistance R_s.

Figure 4.31 Current and voltage waveforms during switching.

$$t_{0n1} = t_{rc} \ln(1 + |V_1/V_2|) \tag{4.94}$$

The charge starts to build up in the base once V_{BE} becomes significantly forward biased. After a time related to the base transit time, the collector current starts to rise. The higher the base current drive, the faster the transistor will turn on. A fair approximation to the second phase of turn-on is:

$$t_{0n2} = t_f \beta_f \tag{4.95}$$

where t_f is the forward delay time (t_{bb} in the simplest case) and β_f is the forced gain I_C/I_{B1} (normally less than the maximum gain for fast turn-on).

The collector current will have reached its final steady-state value in a time given approximately by:

$$t_{0n} = t_{0n1} + t_{0n2} \tag{4.96}$$

Turn-off delay

In many cases the transistor is driven into heavy saturation by a base drive current $I_{B1} = V_1/R_B$, with a corresponding build-up of charge in the base and collector regions, as shown in Fig. 4.32. During turn-off, this charge must be evacuated before the collector current starts to fall. The base current is reversed to a value $I_{B2} \cong V_2/R_B$ as in the diodes of Chapter 3, but the collector current remains constant for a storage delay time t_s given by:

$$t_s = \tau_r \ln[(I_{B2} - I_{B1})/(I_{B2} - I_{Bsat})] \tag{4.97}$$

where $I_{Bsat} = I_C/\beta$ is the base current required to just keep the transistor on the edge of saturation. The time constant τ_r is the inverse delay time of the transistor (analogous to the forward delay time t_f but for the inverse operating mode).

The remaining decay time for the collector current to decay to zero is a rather complex function of the load resistance and collector capacitance time constant, plus the internal delay times of the transistor.

Since all the above time constants for the transistor and also all the circuit resistances and capacitances are modeled in CAD circuit analysis programs such as SPICE, it is not normally necessary to attempt a detailed hand analysis. It is nevertheless useful to know the trends, so as to be able to alter the circuit drive conditions to achieve specific goals, but letting the computer do the job of detailed nonlinear analysis.

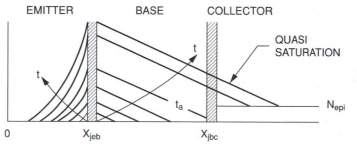

Figure 4.32 Carrier concentration versus distance with increasing time during turn-on of the BJT.

4.9 TEMPERATURE DEPENDENCE OF dc BJT CHARACTERISTICS

The dc collector current is given by Eq. (4.25)

$$I_C = I_{ss} \exp(V_{BE}/V_t) \tag{4.98}$$

where $V_t = kT/q$

$$I_{ss} = (q A D_n / N_{AB} W_b) \tag{4.99}$$

Note that unlike the diode case discussed in Section 3.10, provided we exclude high-current operation, the transistor $I_C(V_{BE})$ characteristic is ideal; that is, there is no factor m. This is due to the fact that space-charge recombination current is supplied by the base contact and is only a component of base and emitter current; it does not affect collector current. We thus have:

$$I_{ss} \propto n_i^2 \propto \exp(-E_g/kT) \tag{4.100}$$

$$I_C = C_1 \exp(-E_g/kT) \exp(q V_{BE}/(kT)) \propto \exp[(-1/kT)(E_g - q V_{BE})] \tag{4.101}$$

Clearly, the result is the same as for the diode with $m = 1$.

Considering the increase in I_C for a small increase in T, say, room temperature to 10°C above room temperature and taking $E_g = 1.1$, $kT(300\text{ K}) = 0.0259$ eV, $V_{BE} = 0.7$ V gives the following results: current increases by a factor 1.7 at $V_{BE} = 0.7$ V, or a factor 2.6 at $V_{BE} = 0.3$ V.

We can thus conclude that the transistor collector current will approximately double for a 10°C rise in temperature around room temperature. As for the diode, it is clearly undesirable in a circuit to establish a fixed V_{BE} bias, because the collector current will be quite unpredictable.

To examine the dependence of BJT voltage on temperature for a fixed collector current, we can rearrange Eq. (4.101) as:

$$-E_g/kT + q V_{BE}/kT = C_2 = \text{ const} \tag{4.102}$$

Differentiating with respect to temperature gives:

$$\frac{E_g}{kT^2} - \frac{q V_{BE}}{kT^2} + \frac{q}{kT} \frac{d V_{BE}}{dT} = 0 \tag{4.103}$$

hence:

$$\frac{d V_{BE}}{dT} = \frac{-E_g}{qT} + \frac{V_{BE}}{T} \tag{4.104}$$

At 300 K this gives: −2.6 mV/°C at $V_{BE} = 0.3$ V, to −1.3 mV/°C at $V_{BE} = 0.7$ V.

Thus depending on the voltage bias the emitter–base voltage decreases by between about 1 and 3 mV/°C at room temperature. This is a very important result, since in most circuit applications the BJT is biased so that the collector current is held approximately constant.

Absolute thermometer

It is worth noting that because of the special property of collector current eliminating all low-current nonideal behavior (space-charge recombination), it is possible to use the BJT as an absolute thermometer. From the Gummel plot shown in Fig. 4.20, the slope of $\log(I_C)$ versus V_{BE} gives the temperature. From Eq. (4.101)

$$\log_{10}(I_C) = \log_{10}(C_1) - [\log_{10}(e)/kT](E_g - qV_{BE}) \tag{4.105}$$

$$\log_{10}(I_{C1}/I_{C2}) = (0.43q/kT)(V_{BE1} - V_{BE2}) \tag{4.106}$$

taking one decade of current, $I_{C1}/I_{C2} = 10$, gives:

$$V_{BE1} - V_{BE2} = 2.33 \ (kT/q) \tag{4.107}$$

By measuring the voltage difference, the absolute value of the temperature may thus be extracted with no calibration necessary.

For example, at 300 K, for one decade of current increase, the voltage changes by $V_{BE1} - V_{BE2} = -60$ mV.

4.10 OVERVIEW OF SPICE BJT CAD MODEL

The Gummel–Poon (GP) CAD model is almost universally used by discrete and integrated circuit chip designers as an aid to development of analogue and digital bipolar and BiCMOS chips. It is embedded in most versions of SPICE [9]. The GP model consists of a set of analytic equations for I–V and $C(V)$ characteristics, plus base and collector resistance components.

The fundamental $I_C = f(V_{BE}, V_{bc})$ equation in the GP model is as follows:

$$I_C = I_{ss}[(\exp(V_{BE}/V_t) - \exp(V_{BC}/V_t)]/q_b \tag{4.108}$$

where

$$I_{ss} = qAD_n n_i^2/(N_{AB}W_b) \tag{4.109}$$

as given by Eq. (4.30). The product $N_{AB}W_b$ has already been discussed and referred to as the base Gummel number. The term q_b is a normalized Gummel number value 1.0 at zero bias.

In the GP SPICE model it describes the effects due to Early voltages and due to high-level injection.

$$q_b = q_1/2 + [(q_1/2)^2 + q_2]^{1/2} \tag{4.110}$$

$$q_1 = 1 + V_{BE}/V_B + V_{BC}/V_A \tag{4.111}$$

$$q_2 = (I_{ss}/I_{KF})[\exp(V_{BE}/V_t) - 1] + (I_{ss}/I_{KR})[\exp(V_{BC}/V_t) - 1] \tag{4.112}$$

where V_A and V_B are the normal and inverse Early voltages.

Use of the I_{KF} parameter to model the I_C versus V_{BE} characteristic is based on the assumption that high-level injection in the quasi-neutral base region is the dominant phenomenon [in which case $I_C \propto \exp(V_{BE}/2V_t)$ at high currents]. The subscript K indicates "knee" current.

An important aspect of the Gummel–Poon model is the representation of high-current f_t fall-off. The model normally employed is based on the Van der Ziel lateral base widening model [7], in which the forward delay time t_f is increased according to:

$$t_{ff} = t_f[1 + X_{TF}(I_{cc}/(I_{cc} + I_{TF})^2 \exp(V_{BC}/1.44V_{tf})] \tag{4.113}$$

This is really an empirical expression with X_{tf}, I_{TF} fitted to model a combination of lateral base widening, Kirk effect (vertical base stretching) and possibly quasi-saturation effects. The V_{tf} term is used to model the high-current V_{CB} dependence.

The circuit analysis program uses these model parameters to compute nonlinear I–V and capacitance voltage values at any dc operating point for small-signal high-frequency analysis, and at any instantaneous values of voltage during a transient analysis. The results are thus more accurate than those possible by hand analysis.

4.11 NOISE PERFORMANCE OF THE BJT

In Appendix N we derive the noise figure of the bipolar transistor.

$$F_n = 1 + r_{bb}/R_g + (I_B/2V_t)(R_g + r_{bb})^2/R_g + (2I_C/V_t)(R_g + r_{bb} + r_\pi)^2/(g_m^2 r_\pi^2 R_g) \tag{4.114}$$

R_g is the source resistance, and all the other terms are as defined already. The noise figure may be minimized by having a small base resistance and operating at moderate base current. If the base resistance is too high, its thermal noise dominates (r_{bb}/R_g). If I_B is too high, the base shot noise term ($\propto I_B$) becomes dominant. If $I_C(= h_{FE}I_B)$ is too low, the last term dominates, since in the denominator $g_m r_\pi = h_{fe}$ and in the numerator $r_\pi \propto 1/I_C$. Noise figures below 3 dB ($F_n = 2$) are common for good transistors with optimized matching and bias conditions. Figure 4.33 shows a typical graph of noise figure versus collector current for different values of source resistance.

4.12 LIMITATIONS OF THE SIMPLE THEORY

The main property of the bipolar transistor that we have not taken into account in the preceding treatment lies in the fact that the real impurity profile is made of two Gaussian-type functions, as was indicated in Fig. 4.2. The calculations for a real BJT involve two modifications:

Noise figure (dB) vs Ic

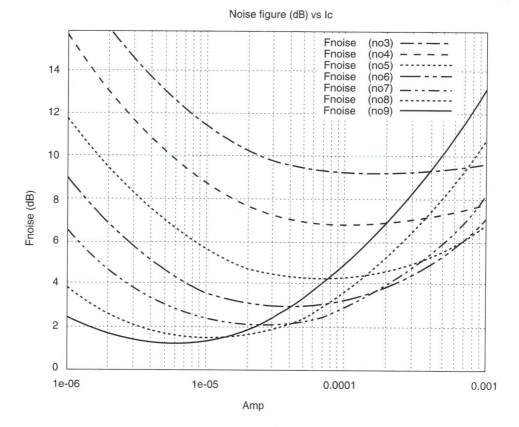

Figure 4.33 Noise figure in dB versus collector current for a transistor with $r_{bb} = 1200$ ohm, $h_{fe} = 22$, $f_t = 4.5$ GHz at 1.5 mA. The source resistance increases on the left side from the top to the bottom curve, the values being: 0.2, 0.4, 1.0, 2.0, 4.0, 10, 20 kohm. Results were generated using the BIPOLE3 simulation program [8].

1. The emitter impurity profile has a steeply graded region next to the junction; this introduces a strong electric field (similar to the NN^+ region of a P^+NN^+ diode), which is in a direction to oppose minority carrier injection. The current gain will thus be enhanced. However, the very high doping levels encountered about halfway toward the surface give rise to additional recombination effects, which makes the carrier lifetime very low. This tends to decrease the current gain [7]. The complete solution can only be obtained using computer simulations.
2. The base region has a doping profile that peaks somewhere between the emitter–base and base–collector junctions as seen in Fig. 4.2 This gives rise to both a retarding electric field (close to the emitter) and an accelerating field (close to the collector). Once again, exact solutions can only be obtained with computer simulations. However for dc properties, the Gummel integral completely characterizes the behavior of minority-carrier flow across the neutral base region; provided the impurity profile is known, the product $N_{AB}W_b$ in Eq. (4.30) can be replaced by the integral: $\int N_A(x)\,dx$, as shown in Appendix O.

In addition, sidewall capacitance and injection from the sidewall regions of the emitter–base junction complicate the determination of the exact electrical behavior, as do the complicated high-current effects (Kirk and quasi-saturation), which we only introduced, without attempting to determine any formulae enabling prediction of operation at high currents. Finally, apart from a brief examination of temperature effects, we completely omitted a study of the very complex thermal behavior due to the fact that the transistor dissipates power, which in turn increases the junction temperature. This can give rise to thermal runaway effects that are an important part of high-power transistor design.

4.13 CONCLUSIONS

This chapter started with an explanation of the vertical *NPN* bipolar transistor structure, including the vertical impurity profile and how it may be approximated for hand analysis to derive simple expressions for terminal currents and current gain. The importance of including the effective emitter doping due to band-gap narrowing was stressed. Operation in the saturation mode was then treated and the ideal "curve tracer" current–voltage characteristics were explained.

Departures from the ideal theory due to high-current and low-current effects were introduced next; these included emitter current crowding, high-level injection, Kirk effect, quasi-saturation, and emitter–base space-charge recombination. The widely used Gummel plot showing the collector and base current dependence on base–emitter voltage was presented. Collector–emitter breakdown was analyzed before proceeding to the important small-signal equivalent circuit of the BJT. This led to derivation of expressions for 3 dB current gain cut-off frequency $f_{c\beta}$ and unity current gain frequency f_t. A detailed calculation for a typical BJT was included. Next, small-signal h parameters were introduced. The integrated circuit BJT including both the vertical *NPN* and lateral *PNP* structures were presented. A brief treatment of switching in the bipolar transistor was followed by an analysis of temperature dependence of the dc characteristics and use of the bipolar transistor as an absolute thermometer. An overview was given of the SPICE Gummel–Poon CAD model and a summary of noise performance of the BJT. The chapter concluded with some limitations of the simple theory.

PROBLEMS

For these problems, the *NPN* BJT has an emitter $10 \times 10 \ \mu m^2$. The base is doped $10^{17} \ cm^{-3}$ over a depth $W_b = 1.0 \ \mu m$ and the corresponding value of $D_n = 20 \ cm^2/s$; the current gain $h_{FE} = 100$; the collector is doped $10^{16} \ cm^{-3}$ with a thickness $W_{epi} = 3 \ \mu m$. For the emitter, assume $N_{Deff} = 10^{18} \ cm^{-3}$, $D_p = 1 \ cm^2/s$.

1. For the given *NPN* bipolar transistor estimate the emitter thickness W_E for a gain $\beta = 100$. Calculate the values of I_{CS}, I_{BS} (include BGN effects).

2. For the given transistor, calculate the collector–base breakdown voltage and the corresponding thickness d_n. Use Eq. (4.46) with $n = 2$ to calculate BV_{CEO} and hence calculate the corresponding value of depletion-layer thickness. State how the given transistor could be improved by changing one vertical thickness.

3. For the given transistor, calculate the ratio of $n_b(0)/n_c$. Hence state the error in calculating the collector current for a given value of V_{BE}. Repeat these calculations if the base width W_b is reduced

to 0.1 μm. What are the implications regarding f_t? (*Hint:* Consider the increase in electron charge in the base.)

4. If the emitter lifetime is 10 ns, calculate the corresponding value of β_{erec}; if the base carrier lifetime is 1 μs, calculate the value of β_{brec}.

5. In order to reduce the base resistance r_{bb}, the base doping level is increased by a factor of 2. Calculate the new value of emitter width W_E required to maintain the same current gain as in the original transistor neglecting emitter recombination. Calculate and sketch current gain versus W_E for an emitter lifetime of 10 ns.

6. For the given transistor, calculate the value of I_C at which emitter current crowding becomes significant ($J_{n(edge)}/J_{n(center)} = 2.7$). If the base doping level is doubled, will this affect the result for I_C? Suggest how the required value of I_C could be increased: (a) without altering the mask dimensions; (b) allowing alterations in mask dimensions.

7. For the above transistor used in the circuit of Fig. 4.5, if $R_L = 10$ kohm, $V_{CC} = 10$ V, calculate the value of I_C, the corresponding base current I_B, and base–emitter voltage V_{BE} at which the transistor just enters saturation. Estimate the value of V_{BE} at which the forced gain $I_C/I_B = \beta_f = 10$.

8. For the above BJT, calculate: (a) the values of V_{BE} and I_C at which high-level injection starts to occur in the base; (b) the value of pinched base sheet resistance R_{BE}/sq and base resistance r_{bb}.

9. (a) Assuming a large value of V_{CE}, calculate the value of I_C corresponding to the onset of vertical base stretching (Kirk effect); (b) if $V_{CE} = 1$ V, calculate the value of I_C at which quasi-saturation starts to occur; (c) sketch β versus I_C for cases (a) and (b).

10. For the given transistor calculate and sketch the electric field versus depth in the epitaxial collector for $n_c = 0, 0.8 \times N_{epi}, N_{epi}, 1.2N_{epi}, 2N_{epi}$. Assume $V_{CB} = 10$ V. Calculate the corresponding values of I_C. Hence determine the value of base width increase W_k for each case and the corresponding reduction in h_{fe} and in peak f_t.

11. Use Eq. (4.54) for the given transistor and calculate the Early voltage V_A and corresponding small-signal output resistance for $I_C = 1$ mA. Assume $V_{CB} = 5$ V.

12. For the above BJT, calculate the depletion-layer thickness d_n for the N-type collector at a reverse bias of 10 V. Hence (one line), calculate the value of d_p on the base side of the c–b junction. Hence calculate the real value of metallurgical base width $X_{jcb} - X_{jeb}$. Using simple ratio calculations, determine the modified values of d_n, d_p if $V_{CB} = 40$ V. Hence find the modified value of W_b at 40 V and the small-signal output resistance (dV_{CE}/dI_C) for a current $I_C = 1$ mA. Compare with the answer to Problem 11.

13. If the base current space-charge recombination term Eqs. (3.36), (3.37) I_{0reb} is 1 μA, estimate the value of V_{BE} at which the $\log(I_B)$ versus V_{BE} plot changes slope. Use this result together with the answers to Problems 8 and 9 to sketch the I_C and I_B Gummel plot for large V_{CE}.

14. Calculate the base and b–c space-charge layer transit times for the above BJT. Hence estimate the peak value of f_t. Calculate C_{je} at $V_{BE} = 0$. If the base diffusion mask dimensions are 15×30 μm^2, calculate C_{jc} for $V_{CB} = 5$ V. Hence sketch carefully f_t and f_{maxosc} versus I_C.

15. For the transistor of Problem 1 using the results of Problem 14, calculate the value of the hybrid π equivalent circuit resistance r_π for $I_C = 1$ mA. Calculate the corresponding f_t and $f_{c\beta}$ frequencies. What is the value of the current gain at 100 MHz?

16. Using the results of Problem 14, for $V_{CB} = 5$ V, calculate the minimum possible ECL propagation delay time.

17. Use the value of C_{je} calculated in Problem 14 to estimate the initial turn-on delay t_{0n1} if the transistor is switched from -5 to $+5$ V through a source resistance of 1 kohm. If the supply voltage V_{CC} is 10 V and the load resistance R_L is 1 kohm, calculate the value of forced gain β_f and the value of t_{0n2}. If after 100 μs the input voltage is switched back to -5 V, and if the inverse delay time is 1 μs, calculate the duration of the storage delay time.

18. For the given transistor, calculate and sketch the noise figure in dB for a source resistance of 300 ohm, as I_C is varied from 1 μA to 10 mA.

19. For the given transistor calculate the factor by which the collector current increases if $V_{BE} = 0.8$ V and the temperature is changed from 27 to 37°C. If the current is kept constant by an appropriate bias network, calculate the decrease in V_{BE} for the same temperature range.

References

1. E. S. Yang, *Microelectronic Devices*. New York: McGraw–Hill, 1988.
2. B. G. Streetman, *Solid State Electronic Devices*. Englewood Cliffs, New Jersey: Prentice-Hall, 1990.
3. A. Bar-Lev, *Semiconductors and Electronic Devices*, 3rd Ed. New York: Prentice-Hall, 1993.
4. D. H. Navon, *Semiconductor Microdevices and Materials*. New York: Holt, Rinehart, Winston, 1986.
5. D. L. Pulfrey, N. G. Tarr, *Introduction to Microelectronic Devices*. Englewood Cliffs, New Jersey: Prentice-Hall, 1989.
6. J. McGregor, D. J. Roulston, et al., *Solid State Electronics* **36**, 391–96, March 1993.
7. D. J. Roulston, *Bipolar Semiconductor Devices*. New York: McGraw–Hill, 1990.
8. D. J. Roulston, *BIPOLE3 Users Manual*. BIPSIM, Inc., and University of Waterloo, June 1996.
9. P. Antognetti, G. Massobrio, eds. *Semiconductor Device Modeling with SPICE*. New York: McGraw–Hill, 1987.

Chapter 5

MOS Field-Effect Transistors

5.1 INTRODUCTION

The MOSFET (metal oxide semiconductor field-effect transistor) is based on the formation of a capacitance whose dielectric is a thin SiO_2 layer on the silicon surface, on top of which is a conducting contact. Originally the contact was metal (Al), but in current technology a heavily doped polysilicon contact is almost exclusively used because of layout advantages. The theory is the same for both structures, and we shall assume a metal contact for simplicity (the technology for the polysilicon gate structure is explained in Chapter 8). In essence, the conducting charge just below the silicon–oxide interface (i.e., at the silicon surface) is controlled by an electric field across the dielectric—hence the name field-effect transistor, or FET. A second FET structure, the junction FET, or JFET, will be described in Chapter 6.

Both the MOSFET and the JFET depend on majority carrier transport; minority carrier injection is only important as a second-order effect. In this respect FET theory is somewhat simpler than the BJT. However, before deriving the MOSFET electrical characteristics, it is necessary to understand the nature of the charge–voltage relations at the silicon surface of the MOS system. Following this, we drive the static current–voltage relations, the small-signal parameters, and the cut-off frequency. Since complementary MOS (CMOS) technology represents the major use of MOSFETs today, the CMOS inverter characteristics are then summarized. Second-order effects become important in advanced VLSI MOSFETs; these relate primarily to short-channel effects and will be studied in some detail. The chapter concludes with a brief summary of MOSFET SPICE models for circuit analysis.

5.2 THE METAL OXIDE SEMICONDUCTOR SYSTEM

Figure 5.1 shows a metal oxide semiconductor (MOS) system. In a MOS transistor the metal forms the *gate* terminal and the semiconductor region is the *substrate*. Because of discontinuities at the crystal surface, creating dangling bonds, and the presence of miscellaneous impurities, a residual surface-state charge exists at the interface between the oxide and the semiconductor. Since in a normal parallel-plate capacitor, the charge at either side of the dielectric can be controlled by the voltage applied between the plates, it is clear that a voltage applied between the metal gate and the substrate allows control of the magnitude and sign of the total charge in the semiconductor near the surface. In order to derive the relations between applied gate voltage and induced charge, it is useful to study the band diagrams for different bias conditions. In Chapter 3 we discussed the band diagrams for a metal–semiconductor system, and we saw how the work function and electron affinity determined the shape of the band diagram (see Section 3.11 and Fig. 3.29). Let us now study the corresponding diagrams for the MOS system under different bias conditions. Throughout this chapter we will focus on MOS structures on P substrates, which as we shall see are used for N-channel MOSFETs. All results for P-channel devices on N substrates may be obtained by direct analogy.

Figure 5.2 shows the situation existing in a Metal–SiO$_2$–Si system with a P substrate, as the bias is varied between extremes. The charge diagrams are shown on the right-hand side. On the left are band diagrams, obtained by considering the work functions of the metal and semiconductor. We recall that in thermal equilibrium in any system, the Fermi levels must align. It is the work function difference plus the surface-state charge Q_{ox} that determine the equilibrium condition shown in the third diagram, where $Q_M = 0$. Q_B is the charge within the depletion layer in the bulk substrate, which must exactly balance Q_{ox} in this case of zero applied bias. There is a residual potential across the oxide, corresponding to the difference in the energy bands. We need not concern ourselves with the details of the band bending, but it is instructive to have a general idea of the relative scenarios as the applied gate substrate bias is varied from a negative voltage to a positive voltage. The reader is encouraged to refer to the band diagrams for the metal–semiconductor system of Chapter 3 (Fig. 3.29).

In the first case, the negative potential on the gate, corresponding to the charge Q_M, induces a positive charge Q_h in the silicon. Charge balance gives $Q_M = Q_{ox} + Q_h$. Since Q_h is larger than the normal majority carrier hole charge, this condition is referred to as *accumulation*.

The second case is included for completeness and is referred to as the *flat-band* condition, where the charge on the metal Q_M just balances Q_{ox}. ϕ'_{MS} is the modified work function difference between the metal and the semiconductor and differs from ϕ_{MS} as discussed for

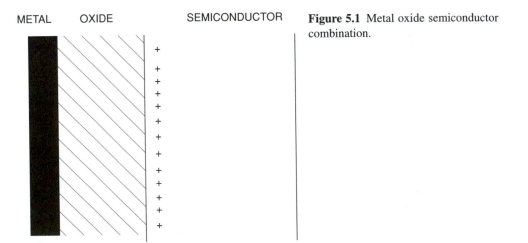

METAL   OXIDE   SEMICONDUCTOR

Q$_{ss}$

Figure 5.1 Metal oxide semiconductor combination.

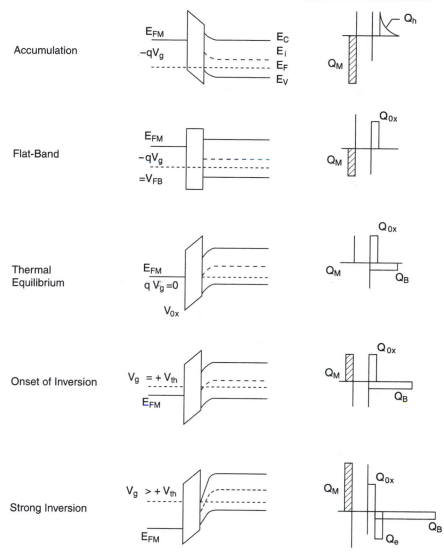

Figure 5.2 Band and charge diagrams for various bias levels in a P substrate MOS system.

the metal–semiconductor system in Chapter 3 because in the MOS system the relevant energy level is the conduction-band edge in the SiO_2 layer.

$$\phi'_{ms} = \phi'_m - \phi'_s \tag{5.1}$$

The metal work function is reduced by 0.9 eV and the semiconductor electron affinity by a similar amount. The flat-band voltage is given by:

$$V_{FB} = \phi'_{MS} - Q_{ox}/C'_{ox} \tag{5.2}$$

where we have introduced the gate capacitance per unit area, C'_{ox}, defined by:

$$C'_{ox} = \epsilon_{ox}/t_{ox} \tag{5.3}$$

ϵ_{ox} is the permittivity of the oxide layer ($= \epsilon_0 \epsilon_{r0x}$) and t_{ox} is the thickness of this layer.

The semiconductor modified work function (see Fig. 3.29) is:

$$\phi'_S = \chi'_s + (E_C - E_i)/q - \phi_F \tag{5.4}$$

where the quantity ϕ_F is the energy difference between the Fermi level and the intrinsic level derived in Appendix F (F.20):

$$\phi_F = V_t \ln(N/n_i) \tag{5.5}$$

where N is equal to N_D for N-type material and N_A for P-type material. Note that we continue to use $V_t = kT/q$. ϕ_F is positive for N-type and negative for P-type material. χ'_s is the modified electron affinity. Equation (5.4) may be rewritten in terms of the band-gap energy E_g as:

$$\phi'_S = \chi'_s + E_g/2q - \phi_F \tag{5.6}$$

Note on notation: Some textbooks express work functions and electron affinity in electron volts (eV), using the same symbols ϕ_{MS}, ϕ_F, χ_s, etc. Throughout, we express these quantities in volts.

Example of ϕ'_{MS} calculation for Al to N-type silicon doped 10^{15} cm^{-3}.

$\phi'_M = 3.2$ V; $\chi'_s = 3.25$ eV; Eq. (5.5) gives $\phi_F = 0.0259 \ln(10^{15}/1.6 \times 10^{10}) = 0.286$ V.

Hence $\phi'_{MS} = 3.2 - (3.25 + 1.12/2 - 0.286) = 3.2 - 3.52 = -0.32$ V.

For P-type material with the same doping level we find:

$$\phi'_{MS} = 3.2 - (3.25 + 1.12/2 + 0.286) = 3.2 - 4.1 = -0.9 \text{ V}.$$

The third case, where $Q_M = 0$, has already been discussed.

In the fourth case, the potential applied to the gate is sufficiently positive that a slight negative charge of electrons is induced at the silicon surface. As this potential is further increased, the negative charge Q_e increases, as shown in the bottom diagram. This situation is referred to as "inversion" since the surface layer now appears to be N type. In the band diagram, the Fermi level has moved close to the conduction-band edge, as in normal N-doped material.

The bias required to "start" the creation of a significant inversion layer is referred to as the threshold voltage V_{th}. *Note that some texts use V_T for the threshold voltage (not to be confused with our use of V_t for kT/q).* It is usually defined as the gate bias required to move the Fermi level at the semiconductor surface to a position corresponding to an equal and opposite doping level (i.e., for a P-type substrate, the surface electron concentration $n = N_A$), that is, to shift it by an amount $2\phi_F$. The threshold voltage is therefore the sum of the flat-band voltage V_{FB}, the inversion layer voltage $2\phi_F$, and the voltage required to support the depletion layer charge Q_B:

$$V_{th} = V_{FB} + 2\phi_F - Q_B/C'_{ox} \tag{5.7}$$

Q_{ox} depends on surface preparation prior to depositing the SiO_2 layer and in modern devices is of order $q \times 10^{11}$ C per cm^2 or less.

Q_B is the charge in the substrate depletion layer given by:

$$Q_B = qN_Ad_p \tag{5.8}$$

where from Eq. (2.62)

$$d_p = \sqrt{[(2\epsilon(V_s + 2\phi_F)]/(qN_A)} \tag{5.9}$$

where ϵ is the permittivity for silicon and V_s is the potential at the silicon surface with respect to the substrate. At the onset of inversion, V_s is zero (no depletion layer). In the basic MOS system V_s reaches a maximum value for strong inversion, when the Fermi level moves up near the conduction-band edge E_c. This maximum value is approximately $2\phi_F$. The charge Q_B varies with V_S as:

$$Q_B/C'_{ox} = (Q_{BO}/C'_{ox})[1 + V_s/(2\phi_F)]^{1/2} \tag{5.10}$$

where Q_{BO} is the charge at the onset of heavy inversion corresponding to the voltage $2\phi_F$:

$$Q_{BO} = (4\,\epsilon q\phi_F N_A)^{1/2} \tag{5.11}$$

and the ratio Q_{BO}/C'_{ox} is referred to as the *body-effect* parameter.

V_{th} is slightly bias dependent because of the Q_B term in Eq. (5.7) [function of V_S in Eq. (5.10)] and may be either positive or negative. The dependence on V_s via Eqs. (5.7)–(5.9) represents the substrate bias dependence of V_{th} in actual MOSFETs.

Example of threshold voltage calculation:

Consider a metal gate silicon NMOS structure with an oxide thickness $t_{ox} = 0.1\,\mu$m (100 nm), $Q_{ss} = 10^{11}$ cm^{-2}, doping level $= 10^{15}$ cm^{-3} E$_{rox} = 3.9$; we shall use E$_{ox} = $ E$_o$ E$_{rox} = 0.3 \times 10^{-12}$ F/cm:

$$Q_{ox}/C'_{ox} = (1.6 \times 10^{-19} \times 10^{11})/(0.3 \times 10^{-12}/0.1 \times 10^{-4}) = 0.53 \text{ V}$$

Using the previous result for N-type material, $\phi'_{MS} = -0.32$ V and the flat-band voltage is thus

$$V_{FB} = -0.32 - 0.53 = -0.85 \text{ V}$$

For P-type material of the same doping level, $V_{FB} = -0.9 - 0.53 = -1.43$ V.

For a substrate doping level $N_A = 10^{15}$ cm^{-3}, $\phi_F = 0.32$ V. Thus under full inversion when $V_S = 2\phi_F$, Eq. (5.9) gives $d_p = 0.9$ μm and $Q_B = 1.44 \times 10^{-8}$ C

Hence $Q_B/C'_{ox} = 1.44 \times 10^{-8}/(0.3 \times 10^{-12}/0.1 \times 10^{-4}) = 0.48$ V.

Thus from Eq. (5.47): $V_{th} = -1.43 - 2 \times 0.286 + 0.48 = -0.38$ V.

For an N-type substrate of the same doping level: $V_{th} = -0.85 + 0.57 - 0.48 = -0.76$ V.

The MOS capacitance

The MOS capacitance C_{MOS} requires some discussion. Referring to the charge diagrams of Fig. 5.2 and neglecting the influence of the surface-state charge Q_{ox}, we can identify two regions of charge that form the input small-signal capacitance: (1) the gate oxide capacitance, C'_{ox}; (2) the depletion-layer capacitance per unit area C'_D given by Eq. (5.12).

$$C'_D = \epsilon/d_p \tag{5.12}$$

where ϵ is the permittivity of the semiconductor ($= \epsilon_r \epsilon_s$). The potential between the silicon surface and the substrate varies with applied gate voltage V_G. For negative gate voltages, referring to Fig. 5.2, the depletion layer collapses, d_p tends to zero and C'_D becomes infinite. As V_G passes through V_{FB}, a small depletion layer forms and C'_D starts to decrease. As V_G is further increased, the substrate potential reaches its maximum value for heavy inversion (the bottom diagram of Fig. 5.2) and V_S becomes approximately constant at the value $2\phi_F$.

The fraction of V_G that is above this value occurs across the oxide layer.

It is clear that the gate oxide and the depletion layer capacitances appear as a series combination, as shown in Fig. 5.3, if measured between gate and substrate.

The total capacitance is thus given by:

$$1/C_{MOS} = 1/C'_{ox} + 1/C'_D \tag{5.13}$$

The overall capacitance–voltage characteristic is therefore as shown in Fig. 5.4.

Measurement of the capacitance is obtained by applying a small rf signal between gate and substrate. At high frequencies, the expected lower curve is obtained. However, *at very low frequencies*, of order 1 kHz, as an inversion layer is formed at high applied bias, the rf current can follow a resistive path along the surface; this is the resistance shown in Fig. 5.3. The depletion-layer capacitance is "shunted," and the input capacitance reverts to that due only to the gate oxide layer. The low-frequency part of the curve starts to increase at $V_{GB} = V_{th}$ (i.e., when the surface inversion layer charge provides a resistive path along the surface).

The situation is further complicated by the presence of surface-state charge Q_{ox} and the work function difference between the metal and the semiconductor. Increasing Q_{ox} results in a lateral shift on the voltage axis of the curves shown in Fig. 5.4, where the flat-band capacitance C_{FB} is the value at $V_G = V_{FB}$. Note that because of the above properties of the MOS capacitance versus voltage, in particular Eqs. (5.9) and (5.12) relating capacitance to doping level (as the

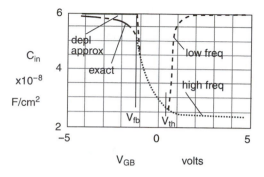

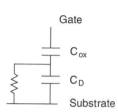

Figure 5.3 MOS oxide capacitance C_{ox} in series with substrate depletion-layer capacitance C_D.

Figure 5.4 MOS capacitance versus gate voltage for $t_{ox} = 500$ Å, $N_A = 2 \times 10^{16}$ cm^{-3}, $Q_{ox} = 10^{11}$ cm^{-2} *(generated by Student BIPOLE with labels added).* (See also CD-ROM DOS or Windows version, type "PLOTCIN MOSYS"; also see effect of varying Q_{ox} by typing "MOSQ" and the effect of varying doping by typing "MOSD.")

doping level increases, the capacitance increases and the right-hand side of the curves move upwards), measurements of C_{MOS} versus V_{GS} may be used to deduce the substrate doping level. The value of the depletion-layer capacitance at V_{FB} is determined by the Debye length (2.63), although the depletion approximation predicts an infinite value (i.e., the total input capacitance would be C_{ox}) (see Tyagi [5]).

5.3 MOSFET STRUCTURE AND *I–V* CHARACTERISTICS

The MOS transistor or MOSFET consists of a MOS capacitance region, the *gate* region, where the surface charge is controlled by the gate bias, and two identical diffused regions at either side, the *source* and *drain* diffusions. Figure 5.5 shows a mask layout and cross section of a complete metal gate *N*-channel MOS transistor on a *P* substrate. "*N* channel" means that the inversion layer is formed by electrons.

The gate dimensions are critical in determination of the MOSFET characteristics. L_{mask} is shown as the horizontal mask distance between source and drain. This is slightly different from the electrical gate length L to be used below. Z is the width perpendicular to L_{mask} of the source and drain diffusions. For this *N*-channel MOSFET the drain is biased at a positive potential with reference to the source (V_{DS} positive). When an inversion layer is formed (by applying a sufficiently positive gate bias V_{GS}), electrons flow from source to drain; the drain current I_{DS} is positive. In this metal gate MOSFET, the gate metal must overlap substantially the source and drain diffusions to ensure that the inversion layer is continuous between source and drain. Even in a self-aligned polysilicon gate MOSFET (described in Chapter 8), the sideways spread of the source and drain implants/diffusions creates a similar scenario (but with minimal overlap).

A more detailed cross section of the important region of the MOS transistor is shown in Fig. 5.6. For a given V_{GS} and V_{DS}, the inversion-layer charge will decrease from left to right

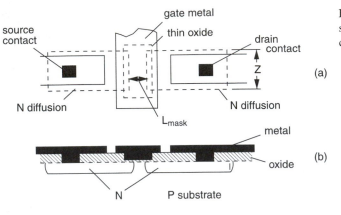

Figure 5.5 Metal gate MOSFET structure: (a) mask layout; (b) cross section.

because of the voltage drop in the conducting layer. In the following treatment the channel length L is defined from the edge of the source diffusion to the point where the value of Q_e becomes zero. This is less than the source–drain diffusion spacing by an amount w_{scl}, the drain depletion layer thickness in the horizontal direction at the surface. Note that L is less than the corresponding mask dimension L_{mask} by the sum of the source and drain lateral diffusions, $2r_j$, plus w_{scl}. In other words,

$$L = L_{mask} - w_{scl} - 2r_j \tag{5.14}$$

Beyond $y = L$ the electrons travel at the saturated drift velocity in the high-field depletion layer region as in a bipolar transistor collector space-charge layer. The following analysis is based on the Shockley gradual channel approximation, with constant channel mobility in the neutral conducting channel.

The relation between the (bias-dependent) surface charge per unit area $Q_\square$ and the field E_{ox} across the oxide layer is given by Gauss's law:

$$\epsilon_{ox} E_{ox} = Q_\square \tag{5.15}$$

The vertical field E_{ox} is given by the total voltage at any point y along the channel:

$$E_{ox} = V_{tot}(y)/t_{ox} \tag{5.16}$$

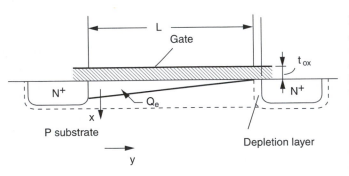

Figure 5.6 Cross section of N-channel MOS transistor. Q_e is the inversion layer electron charge; the broken line represents the depletion layer boundary. The distance L is slightly less than the source-to-drain diffusion distance by an amount equal to the horizontal drain depletion layer width w_{scl}.

We thus have a relation between the surface charge of electrons $Q_e(y)$ and the potential:

$$Q_e(y) = C'_{ox} V_{tot}(y) \qquad (5.17a)$$

or

$$Q_e(y) = C'_{ox}[(V_{GS} - V_{th}) - V(y)] \qquad (5.17b)$$

where we have used $C'_{ox} = \epsilon_{ox}/t_{ox}$ [Eq. (5.3)] and where V_{GS} is the applied gate–source bias and $V(y)$ the voltage at any point y with the source taken as $V(0) = 0$. We assume V_{th} to be constant in the following analysis.

The drain–source drift current (positive in the drain-to-source direction with V_{DS} positive and the horizontal electric field $E_y(y)$ negative; electrons flow from left to right in Fig. 5.6) may be written in terms of $E_y(y)$ and electron concentration per cubic centimeter $n(y)$ as:

$$I_{DS} = -q\mu_e n(y) A E_y(y) \qquad (5.18)$$

where A is the cross-sectional area of electron charge, $n(y)$, and μ_e is the electron mobility of the inversion-layer electrons. Because of additional scattering at the surface, this is considerably lower than the bulk mobility given in Fig. 2.11. A typical value for the surface layer μ_e is 600 cm²/V s (instead of over 1000 cm²/V s in lightly doped bulk silicon).

Since $n(y)$ extends an unknown distance into the substrate, we should rewrite Eq. (5.18) thus:

$$I_{DS} = -q\mu_e Z E_y(y) \int_0^\infty n(y)\, dx \qquad (5.19)$$

where x is the distance perpendicular to the surface. The integral of $n(y)\, dx$ is simply the concentration of electrons per square cm. This is directly related to the charge per square centimeter. Note that we do not need to know the shape or the extent of penetration of this charge layer, although numerical simulation indicates a penetration below the surface of several atomic layers, typically less than 10 nm.

$$Q_e(y) = -q \int_0^\infty n(y)\, dx \qquad (5.20)$$

We may thus rewrite Eq. (5.19) in the form:

$$I_{DS} = q\mu_e Z E_y(y) Q_e(y)/q \qquad (5.21)$$

Substituting $E(y) = -dV(y)/dy$ gives:

$$I_{DS} = -\mu_e Z Q_e(y) \frac{dV(y)}{dy} \qquad (5.22)$$

[note that $Q_e(y)$ is negative and hence I_{DS} is positive]

Integrating from $y = 0$ to L, corresponding to $V(y) = 0$ to $V(y) = V(L) = V_{DS}$, gives:

$$\int_0^L I_{DS}\, dy = -\mu_e Z \int_0^{V_{DS}} Q_e(y)\, dV(y) \qquad (5.23)$$

or

$$I_{DS}L = \mu_e Z C'_{ox} \int_0^{V_{DS}} [(V_{GS} - V_{th}) - V(y)] \, dV(y) \qquad (5.24)$$

$$= \mu_e Z C'_{ox}[(V_{GS} - V_{th})V_{DS} - V_{DS}^2/2]$$

Hence we obtain the final result for the MOSFET $I_{DS} = f(V_{GS}, V_{DS})$ characteristics:

$$I_{DS} = \mu_e(Z/L)C'_{ox}[(V_{GS} - V_{th})V_{DS} - V_{DS}^2/2] \qquad (5.25)$$

Note that this equation is only valid for the range:

$$V_{DS} \leq V_{GS} - V_{th} \qquad (5.26)$$

If used for V_{DS} greater than $V_{GS} - V_{th}$, the I_{DS} curves will decrease toward zero. This is a nonphysical result; so care must be taken in use of the above equation.

The threshold voltage is considered constant in the above derivation. In fact, the horizontal change in surface potential V_S due to the drift current I_{DS} causes a change in Q_B and hence from Eq. (5.7) a change in V_{th} along the channel. The ideal I_{DS}–V_{DS} characteristic will therefore be slightly modified.

Figure 5.7 shows the $I_{DS} = f(V_{DS}, V_{GS})$ characteristics obtained from this equation. For V_{DS} values greater than the pinch-off value given by Eq. (5.26), I_{DS} is assumed to be constant. In reality there is a finite gradient due to second-order effects related to a decrease in effective channel length with increasing V_{DS}. Avalanche multiplication will also eventually cause an increase in current, leading to eventual breakdown as in the BJT.

The "pinch-off" locus given by Eq. (5.26), $V_{DS} = V_{GS} - V_{th}$ is indicated by the broken curve. The region to the right is referred to as the *saturation region*, and to the left it is sometimes called the *triode* region. A linear or ohmic region is also defined for low V_{DS} values near the origin. Note the difference in notation with the BJT, where *saturation* has a completely different meaning.

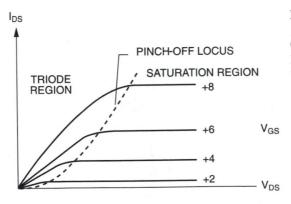

Figure 5.7 Ideal MOSFET $I_{DS} = f(V_{DS}, V_{GS})$ characteristics. (See also CD-ROM Windows BIPGRAPH example MOS1; "I_{DS} vs V_{DS}" [in DOS version, type "PLOTIVD MOS1"].)

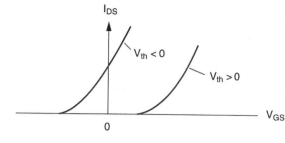

Figure 5.8 $I_{DS} = f(V_{GS})$ characteristics in saturation ($V_{DS} > V_{GS} - V_{th}$) for a depletion and an enhancement N-channel MOSFET. (See also CD-ROM Windows BIPGRAPH examples MOS1 and MOS3 together; "I_{DS} vs V_{GS}." [In DOS version, type "PLOTIVG MOS1" and "PLOTIVG MOS3"].)

The Shockley theory assumes that the current I_{DS} remains constant for V_{DS} greater than the pinch-off value. To determine the I_{DS} versus V_{GS} characteristics in the saturation region, therefore, we set $V_{DS} = V_{GS} - V_{th}$ in Eq. (5.25). This gives:

$$I_{DS(pinchoff)} = \mu_e(Z/L)C'_{ox}(V_{GS} - V_{th})^2/2 \qquad (5.27)$$

The I_{DS} versus V_{GS} characteristic given by Eq. (5.27) is plotted in Fig. 5.8 for two typical devices. In one case the threshold voltage is negative and current flows even when $V_{GS} = 0$. This is called a *depletion* MOSFET. In the other, more common, case, V_{th} is positive, and the device is referred to as an *enhancement* MOSFET.

5.4 SMALL-SIGNAL CHARACTERISTICS

5.4.1 Low-frequency parameters

Transconductance

The transconductance g_m is of interest only in the saturation region; so we use Eq. (5.27):

$$g_m = \frac{dI_{DS(pinchoff)}}{dV_{GS}} = \mu_e(Z/L)C'_{ox}(V_{GS} - V_{th}) \qquad (5.28)$$

Drain–source resistance at low V$_{DS}$

We may use Eq. (5.27) again to determine an expression for $1/r_{on}$, the gradient dI_{DS}/dV_{DS} at low V_{DS}. Setting V_{DS} to a low value, the V_{DS}^2 term becomes negligible compared to the linear term. Thus we obtain:

$$\frac{1}{r_{on}} = \frac{dI_{DS}}{dV_{DS}} = \mu_e(Z/L)C_{ox}(V_{GS} - V_{th}) \qquad (5.29)$$

This appears to be the same as the expression for g_m given by Eq. (5.28). It is very important to note, however, that Eq. (5.27) is valid only in saturation, whereas Eq. (5.29) applies only at very low V_{DS}, in the linear I_{DS}–V_{DS} region.

Example of basic MOS calculation:
Consider an N-channel MOSFET with channel length $L = 20\,\mu m$, width $Z = 4\,\mu m$, oxide

thickness $t_{ox} = 0.1 \, \mu m$, threshold voltage $V_{th} = 2$ V. Assume electron mobility $\mu_e = 500 \, cm^2/V \, s$. Find: (1) I_{DS} and g_m for $V_{GS} = 6$ V, $V_{DS} = 6$ V; (2) r_{on} for $V_{GS} = 6$ V, $V_{DS} = 1$ V.

$$\mu_e(Z/L)C'_{ox} = 500(4/20) \times 0.3 \times 10^{-12}/0.1 \times 10^{-4} = 3 \times 10^{-5} A/V^2$$

$V_{DS} = 6$ V is greater than the pinch-off value given by Eq. (5.26) as $6 - 2 = 4$ V. The MOSFET is therefore in the constant current saturation region.

From Eq. (5.27) drain–source current I_{DS} for $V_{GS} = 6$ V is $3 \times 10^{-5} \times (6-2)^2/2 = 0.24$ mA. The transconductance from Eq. (5.28) is $g_m = 3 \times 10^{-5}(6-2) = 0.12$ mA/V.

For $V_{DS} = 1$ V, $V_{GS} = 6$ V, the MOSFET is in the ohmic region and the "on" resistance using Eq. (5.29) is $r_{on} = 1/[(3 \times 10^{-5} \times (6-2)] = 1/1.2 \times 10^{-4} = 8330$ ohm.

Channel length modulation and output conductance

As indicated in Fig. 5.6 and Eq. (5.14), the depletion layer extends laterally from the drain diffusion by an amount w_{scl}. The channel length L is reduced by an amount ΔL as V_{DS} is increased. Since $I_{DS} \propto 1/L$, the drain current increases as $L/(L - \Delta L)$. Our P–N junction result (2.62) may be used to determine approximately the magnitude of this effect by noting that the horizontal depletion-layer thickness may be written as $w_{scl} = w_{sco}(V_{jtot}/V_{bi})^{1/2}$ and that the change ΔL in L is equal to w_{scl}. The small-signal output conductance in the saturation region, g_{out} is given by:

$$g_{out} = \frac{dI_{DS}}{dV_{DS}} = \frac{dI_{DS}}{dL}\frac{dL}{dV_{DS}} \tag{5.30}$$

From Eq. (5.25) at fixed V_{GS} considering only the saturation region of operation, we have:

$$\frac{dI_{DS}}{dL} = -I_{DS}/L \tag{5.31}$$

From Eq. (2.62), using $\Delta L = \Delta w_{scl}$, where w_{scl} is the depletion-layer thickness (instead of d_n) and assuming that $V_{DS} \gg V_{bi}$, we obtain:

$$\frac{dL}{dV_{DS}} = -\tfrac{1}{2}w_{scl}/V_{DS} \tag{5.32}$$

Hence we obtain:

$$g_{out} = (I_{DS}/V_{DS})\tfrac{1}{2}w_{scl}/L \tag{5.33}$$

In order to have a (desirable) small output conductance, it is clear that the depletion-layer thickness w_{scl} must not occupy a significant fraction of the channel length L.

The above result may also be expressed in terms of a parameter V_A, the "MOSFET Early voltage," analaogous to that for the BJT in Chapter 4, Fig. 4.22. From the above result, we may write:

$$g_{out} = I_{DS}/(V_A + V_{DS}) \tag{5.34}$$

Rearranging terms and substituting Eq. (5.33) gives:

$$V_A/V_{DS} = (2L/w_{scl}) - 1 \qquad (5.35)$$

where $w_{scl} \approx w_{sco}(V_{DS}/V_{bi})^{1/2}$

Example of output conductance:
Using previous values, assume a channel doping $N_A = 10^{16}$ cm^{-3} and $V_{DS} = 6$ V;
Eq. (2.62) gives $w_{scl} = 0.92\,\mu$m; hence Eq. (5.33) gives
$g_{out} = (2.4 \times 10^{-3}/6) \times (1/2) \times 0.92/20 = 0.9 \times 10^{-5}$.
Hence $r_{out} = 110$ kohm.
From Eq. (5.35) $V_A/V_{DS} = (20 \times 2/0.92) - 1 = 42.4$. Hence $V_A = 42.4 \times 6 = 254$ V.

Clearly it is the ratio w_{scl}/L that determines the magnitude of this effect, and it is only when the depletion-layer width of the drain–substrate junction in the horizontal direction becomes a significant fraction of the channel length L that the output conductance becomes very high.

Figure 5.9 illustrates the finite slope in the I_{DS}–V_{DS} characteristics due to channel length modulation (the MOSFET Early voltage effect), which becomes more pronounced for short-channel devices. At high V_{DS} values avalanche multiplication and breakdown will occur, determined by the background doping level. This is similar to the base–collector breakdown in a BJT but complicated by the two-dimensional nature of the depletion layer extending from the drain–substrate junction both vertically and laterally.

5.4.2 Small-signal equivalent circuit and f_t

The small-signal equivalent circuit of the MOSFET operating in the saturation region is shown in Figure 5.10. C_{sb} and C_{db} are the junction depletion-layer capacitances of the source and drain diffusions (see Section 3.8.1). For the case where the source and substrate are both grounded C_{sb} is shorted.

C_{gb} is the parasitic capacitance of the gate metallization that extends over the thick (field) oxide region to connect to the rest of the circuit and may be quite significant (although the field oxide is thicker than the gate oxide, the overall metal interconnect can be an order of magnitude greater than the area of the gate).

The capacitance C_{gs} is a function of the oxide and bulk depletion-layer components, but it is normally comparable to $C_{ox} = LZC'_{ox}$.

The gate-to-drain capacitance C_{gd} is due to fringing effects between the gate metal and the drain diffusion and is typically considerably smaller than C_{gs}, but nevertheless not to be neglected because of the feedback effect coupled with the gain of the stage in amplifier applications (known as the Miller effect in circuit theory).

The transconductance g_m has already been discussed [Eq. (5.28)]. The input resistance r_{gs} is due to the resistance of the inversion layer. Its value will be close to r_{on} given by Eq. (5.29).

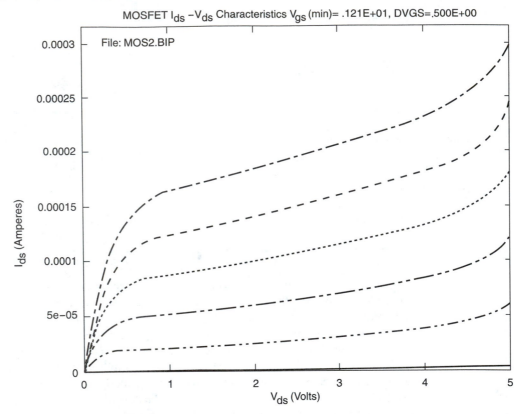

Figure 5.9 $I_{DS}-V_{DS}$ characteristics of a short-channel MOSFET illustrating increased small-signal output conductance at high V_{DS} bias (generated using Student BIPOLE). (See also CD-ROM BIPGRAPH example MOS2; "I_{DS} vs V_{DS}.")

The f_t of the MOSFET may be derived in the same manner as for the JFET. The output current is given by:

$$i_{out} = g_m v_{ox} \qquad (5.36)$$

where v_{ox} is the small high-frequency voltage appearing across the gate oxide capacitance (note that it this voltage that creates the modulation of the channel inversion layer and hence controls the output current). The input current may be written as (neglecting C_{dg} and C_{gb}):

$$i_{in} = v_{ox}\omega C_{ox} \qquad (5.37)$$

The current gain is thus:

$$A_i = i_{out}/i_{in} = g_m/\omega C_{ox} \qquad (5.38)$$

and the unity current gain frequency is thus:

$$f_t = g_m/2\pi C_{ox} \qquad (5.39)$$

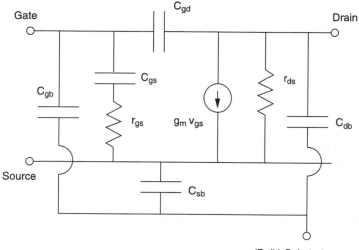

Figure 5.10 Small-signal equivalent circuit of MOSFET in saturation.

A general charge relation for FETs

A general relation for the transconductance of MOSFETs may be obtained by writing:

$$g_m = \frac{dI_{DS}}{dV_{GS}} = \frac{dI_{DS}}{dQ_e}\frac{dQ_e}{dV_{GS}} \tag{5.40}$$

where dQ_e is the incremental change in majority carrier charge in the conducting channel due to an incremental change dV_{GS} in V_{GS}. It is important to note that an increase in conducting channel charge by an amount dQ_e corresponds to an identical decrease in applied gate charge dQ_{GS}. Equation (5.40) can thus be rewritten as:

$$g_m = \frac{dI_{DS}}{dQ_e}\frac{dQ_{GS}}{dV_{GS}} \tag{5.41}$$

We can now identify the two terms in parentheses. The reciprocal of the first term, dQ_e/dI_{DS}, is the channel transit time t_{sd}. The second term, dQ_{GS}/dV_{GS}, is the gate capacitance C_{ox}. We thus obtain the general relation:

$$g_m = C_{ox}/t_{sd} \tag{5.42}$$

It is therefore also possible to express the result for f_t in terms of the source-to-drain transit time:

$$f_t = 1/2\pi t_{sd} \tag{5.43}$$

For the above Shockley theory we can write: $t_{sd} = L/v_{av}$, where $v_{av} = \mu E_{av}$ is the average drift velocity for an electric field E_{av}. Since E_{av} is given approximately by $E_{av} = V_{DSP}/L$ if V_{DSP} is close to the pinch-off value of V_{DS}, this may be put in the form:

$$t_{sd} = L^2/\mu V_{DSP} \tag{5.44}$$

---

f_t calculation for MOSFETs:

Using the values in the previous example, $g_m = 0.12$ mA/V,

$$C_{ox} = LZ \times 0.3 \times 10^{-12}/0.1 \times 10^{-4} = LZ \times 3 \times 10^{-8} = 20 \times 4 \times 10^{-8} \times 3 \times 10^{-8} = 24\,\text{fF}$$

Using Eq. (5.39) gives $f_t = 800$ MHz.

5.5 SWITCHING SPEED OF MOSFETs

Since digital circuits form one of the main applications of the MOSFET, let us determine a simple approximate expression for the switching time in the inverter circuit shown in Fig. 5.11. Consider the inverter initially with an input voltage of zero; the MOSFET is hence in the "off" state since the input is below the threshold voltage; see Fig. 5.8. The output voltage is at the supply voltage V_{DD} (no current, hence no voltage drop in the load resistance R_L). If the input is now suddenly switched to a "high" voltage, which we will assume to be V_{DD}, the output current from the drain terminal is:

$$i_{out} = g_m V_{ox} = g_m V_{in} = g_m V_{DD} \tag{5.45}$$

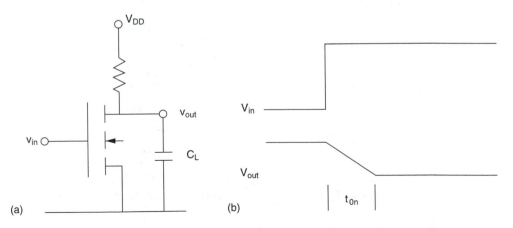

Figure 5.11 (a) MOSFET inverter with capacitive load C_L; (b) ideal switching waveform.

where we approximate the input capacitance as $C_{GS} = C_{ox}$. For most of the "turn-on" transient, this current will be used to reduce the charge Q_c on the load capacitor C_L. Initially C_L was charged to the voltage V_{DD}, and hence at the start of the transient we have:

$$Q_c(0) = C_L V_{DD} \tag{5.46}$$

At the end of the transient the voltage across C_L is low. In a typical inverter of the sort shown in Fig. 5.11, the logic low voltage will be of order 0.5 V. In a CMOS inverter, it will be zero. Approximating the condition that the current remains constant during the turn-on transient, setting $Q_c(t_{on}) = 0$, and defining the duration of this transient to be t_{on}, we have:

$$Q_c(0) - Q_c(t_{on}) = C_L V_{DD} = i_{out} t_{on} \tag{5.47}$$

Using Eq. (5.45), the value of t_{on} is thus:

$$t_{on} = (C_L/g_m) \tag{5.48}$$

It is interesting to consider the special case where the MOSFET inverter is driving an identical inverter. In this case, neglecting parasitic and junction capacitances to determine the absolute minimum value of t_{on}, C_L will be the input capacitance C_{gs} of the following stage, and Eq. (5.48) becomes:

$$t_{on(min)} = C_{ox}/g_m \tag{5.49a}$$

Comparing this result with Eq. (5.39) and (5.42), we see that the turn-on delay time is:

$$t_{on(min)} = t_{sd} \tag{5.49b}$$

This is a very significant result, since it represents the minimum possible turn-on time of a MOSFET inverter driving an identical inverter.

Example of MOS inverter turn-on delay:

Consider the values in the previous two examples with the NMOS transistor being the driver and a load capacitance of 1 pF. Equation (5.48) gives:

$$t_{on} = 10^{-12}/1.2 \times 10^{-4} = 8.3 \times 10^{-9} = 8.3 \text{ ns}$$

If the load capacitance was due only to an identical MOS transistor and if there was no parasitic interconnect capacitance, then $C_L = C_{ox} = 24$ fF, and Eq. (5.49) gives:

$$t_{on(min)} = 24 \times 10^{-15}/1.2 \times 10^{-4} = 200 \times 10^{-12} = 0.2 \text{ ns}$$

The "turn-off" delay time for the above inverter turns out to much longer; this is because if the input voltage is now reduced abruptly from V_{DD} to zero, the load is no longer being charged by the current $g_m(-V_{DD})$ but by a much smaller current. Fortunately this case is largely of only academic interest since most inverters today are CMOS inverters, which we will now discuss.

5.6 THE CMOS INVERTER

5.6.1 The basic structure and characteristics

Figure 5.12 shows the circuit of a CMOS inverter. The bottom MOSFET T_D is an N-channel device; the top MOSFET T_L is a P-channel MOSFET. Because of the common input gate, these devices operate as "mirror images" of each other.

For V_{in} less than the N-channel threshold voltage, T_D is "off," and only a small leakage current will flow. In this condition, the gate–source voltage of the load MOSFET T_L is ($V_{in}-V_{DD}$), which will be such as to keep T_L in its "on" condition with a drain–source resistance given by Eq. (5.29), which as we saw above is typically of order kohm. This is in series with the extremely high drain–source resistance of T_D; so the output voltage v_{out} will be equal to V_{DD}. At the other extreme, when v_{in} is high, close to V_{DD} such that the magnitude of $V_{DD}-v_{in}$ is less than the threshold voltage of T_L, the opposite situation exists; T_D is now "on," T_L is "off," and $v_{out} = 0$. We have thus an ideal situation for a logic inverter: a voltage swing equal to the supply voltage together with zero current in both logic low and logic high states. Figure 5.13 shows the dc transfer characteristic.

It is desirable to have similar characteristics for the P- and N-channel devices (so that the input voltage steps are the same in magnitude when switching from on to off as from off to on states. The threshold voltages may be adjusted by controlling the channel doping level [see Eq. (5.7), where the Q_B term depends on doping level via Eq. (5.18) and where ϕ_s' (5.6) and ϕ_F (5.5)

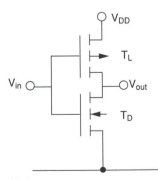

Figure 5.12 CMOS inverter with a P-channel device T_L as the top (load) MOSFET and an N-channel device T_D as the bottom (driver) MOSFET.

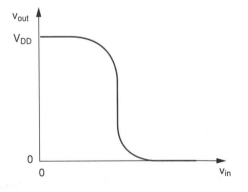

Figure 5.13 dc transfer characteristic of CMOS inverter.

both involve the doping level under the gate oxide]. This adjustment is usually accomplished by ion implantation. If the implanted layer is thin, with a total number of implanted impurity atoms called the *dose* D_I, the change in threshold voltage ΔV_{th} is simply the modified Q_B term in Eq. (5.7):

$$\Delta V_{th} = q D_I / C'_{ox} \tag{5.50}$$

Although the channel mobility of the *N*- and *P*-channel devices differs by approximately a factor of two, the $I_{DS}-V_{GS}$ characteristics may be made identical by suitable alteration of the L/Z ratios for the two MOSFETs [see Eq. (5.25)]. The turn-on and turn-off switching times will also be comparable under these conditions.

Figure 5.14 shows the essential features of a CMOS inverter for an integrated circuit, using one *N* well for the *P*-channel device.

Apart from threshold voltage adjustment, the outstanding design problem is to prevent a phenomenon known as *latchup*. The horizontal path in the above diagram near the surface "*abcd*" is in fact a *P–N–P–N* device. The dc characteristic of a *PNPN* device (a Shockley diode or SCR structure) will be discussed in Chapter 7 and is shown in Fig. 5.15.

The possibility of switching to the stable I_H holding current condition exists during a transient, if the condition $\alpha_{PNP} + \alpha_{NPN} = 1$ is satisfied, where α is the common-base current gain $[= \beta/(1 + \beta)]$ of the lateral BJT. This is referred to as CMOS *latchup*. Such a condition has the potential of high-current destruction, or of destroying the logic operation of the circuit, and must be avoided. In order to reduce the lateral gain to a condition where $\alpha_{PNP} + \alpha_{NPN} < 1$, the BJT base region can be highly doped by adding an extra N^+ diffusion at *b* in Fig. 5.14 (or alternatively an extra P^+ diffusion at *c*). These diffusions have the drawback of adding a high parasitic junction capacitance. A preferred technique is therefore to use oxide isolation

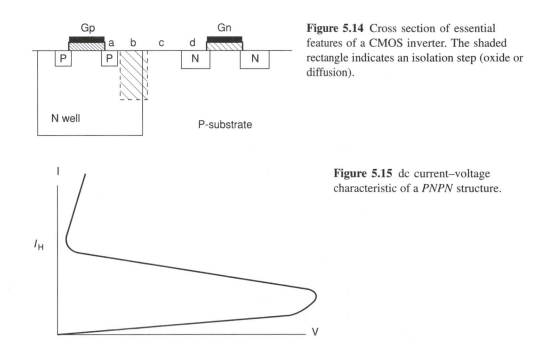

Figure 5.14 Cross section of essential features of a CMOS inverter. The shaded rectangle indicates an isolation step (oxide or diffusion).

Figure 5.15 dc current–voltage characteristic of a *PNPN* structure.

(see Section 8.7.2) for the shaded area. Not only does this eliminate the lateral base of the BJT, but it also eliminates the sidewall capacitance of the existing shallow diffusions, thereby decreasing the overall parasitic capacitance.

5.6.2 Switching times of CMOS inverters

In switching the CMOS inverter, turn-off for the NMOSFET will correspond to turn-on for the PMOSFET. Hence if both devices are made with the same transconductance and if each has the same capacitance, the turn-on and turn-off delays of the CMOS inverter will be similar, with each given by Eq. (5.48). Note, however, that since the mobility of the P- and N-channel devices differ by a factor of about 2, to have symmetrical dc transfer characteristics, the L/Z ratios will be different. It is important to note also that the total load capacitance will be given by:

$$C_L = C_{gs(N)} + C_{gs(P)} + C_{db(N)} + C_{sb(P)} + C_{metal} \qquad (5.51)$$

where the added subscripts refer to the N- and P-channel devices. C_{metal} is the capacitance of the metal interconnect between the driver inverter and the twin gates of the following stage.

The junction capacitances may be calculated from Eqs. (2.62) and (3.45). The result of Eq. (2.62) tells us that for a background doping level of 10^{16} cm^{-3}, the depletion-layer thickness is approximately 0.3 μm at zero bias. Since the dielectric constant of SiO$_2$ is approximately one-third that of silicon, this means that for equal source diffusion, drain diffusion, and gate areas (a simple condition chosen only to get a feel for the magnitudes), and for an oxide thickness of 0.1 μm, the total load capacitance will be approximately $4C_{gs}$. Metal interconnect capacitance can easily double this value. We may therefore conclude that the switching delay time of a pair of CMOS inverters will be typically at least an order of magnitude larger than the value given by the source–drain transit time t_{sd}.

5.7 SECOND-ORDER EFFECTS IN MOSFETS

In order to increase the packing density of silicon chips and at the same time to improve switching speed, dimensions of CMOS structures are kept as small as possible consistent with lithography limitations. For channel lengths less than about 1 μm, some new physical effects become important. In the following sections a detailed study of inversion-layer charge will first be presented (although this theory is valid for all MOSFETs, it becomes particularly important in short-channel devices). Then several important short-channel effects will be briefly discussed.

5.7.1 Inversion-layer charge

In the preceding treatment we implicitly assumed that the electron inversion layer charge Q_e was zero for V_{GS} less than V_{th} and then became linearly proportional to V_{tot} as in Eq. (5.17). In fact, the inversion layer is formed gradually, as implied by the scenarios depicted in the last two diagrams of Fig. 5.2. As the surface potential V_S increases, the band gradually bends so that the

Fermi level E_F rises above the intrinsic level E_i. At this point the surface electron concentration would be equal to n_i, and no significant conduction occurs. The dependence of Q_e on V_S may be solved by using Gauss's law to relate the total charge $Q_S = Q_e + Q_B$ and surface field E_S and Poisson's equation to relate the space charge in the depletion layer to the potential across this layer (including only the ionized acceptor charge qN_A and the free-electron charge qn). The analysis is given in Appendix P and leads to the result for inversion-layer charge:

$$Q_e = -V_t \left(\frac{2q\epsilon N_A}{V_t} \right)^{1/2} \left[\left(V_S/V_t + (e^{V_S/V_t} - 1)e^{-2\phi_F/V_t} \right)^{1/2} - \left(V_S/V_t \right)^{1/2} \right] \quad (5.52)$$

The dependence on gate bias V_G is also obtained in Appendix P as:

$$V_G = V_S + V_{FB} + Q_S/C_{ox} \quad (5.53)$$

where $Q_S = Q_e + Q_B$ and $Q_B = -(2\epsilon\,qN_A V_S)^{1/2}$.

Typical results are plotted in Fig. 5.16. It is clear that although Q_e increases initially exponentially with V_G once the V_{FB} value has been exceeded, it only becomes significant for V_G greater than the threshold voltage V_{th} and then increases linearly as (V_G-V_{th}). It may help to get magnitudes into perspective by recognizing that a charge concentration per square centimeter $Q_e/q = 10^{12}$ cm^{-2} corresponds to an electron concentration of 10^{18} cm^{-3} and 10^{14} cm^{-2} corresponds to 10^{21} cm^{-3}. It is clear from this figure that the approximation of $Q_e = 0$ for $V_G < V_{th}$ and of Q_e proportional to (V_G-V_{th}) for V_G greater than V_{th} by a volt or so is excellent for predicting I–V characteristics over the useful range of operation of a MOSFET.

Note also that for $V_G \gg V_{th}$, the surface potential "saturates" at a value slightly above $2\phi_F$ as the Fermi level moves close to the conduction-band edge.

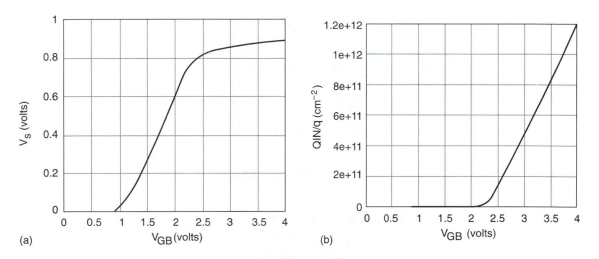

(a)   (b)

Figure 5.16 Subthreshold behavior: (a) surface voltage versus gate voltage V_G; (b) inversion-layer charge ($QIN = Q_e/q$ in units of cm^{-2}) versus V_{GB}. The MOSFET details are: $V_{FB} = 0.867$ V, $\phi_F = 0.370$ V ($N_A = 2 \times 10^{16}$ cm^{-3}), $t_{ox} = 25$ nm, $V_{th} = 2.20$ V. Results generated using Student BIPOLE. (See also CD-ROM DOS or Windows version, type: PLOTVS MOSYS and PLOTQI MOSYS.)

5.7.2 Subthreshold operation

In some cases it is desirable to consider the operation of the MOSFET slightly below the threshold voltage. For $V_{FB} < V_G < V_{th}$ operation is said to be in the subthreshold region. This can give rise to a source–drain diffusion current neglected in the preceding formulation. We can write this current in the form:

$$I_{DSdiff} = (qZD_n/L)[Q_e(L) - Q_e(0)] \tag{5.54}$$

where the integral is over all x. From Eq. (5.52) it is clear that due to the $\exp(V_S/V_t)$ term if the drain and source potentials are different, even a small difference between V_{SB} and V_{DB} can create a significant charge gradient. This give rise to a diffusion current I_{DS} where [1]:

$$I_{DS} = K_{VS}[\exp(-V_{SB}/V_t) - \exp(-V_{DB}/V_t)] \tag{5.55}$$

where K_{VS} is a constant (proportional to $1/L$) for a given V_S value (given V_G). This current is similar to the collector current for a bipolar transistor. In this case the source is an N-type emitter, the "channel region" P-type, the drain a reverse biased N-type collector. This diffusion current is negligible for normal MOSFET operation, where V_{GS} is greater than V_{th}, but can become greater than the normal drift current expressed by Eq. (5.25) in the subthreshold region for short-channel MOSFETs. This effect is further enhanced when the source and drain depletion layers almost meet (within a few extrinsic Debye lengths, L_D, of order 0.1 μm). Then the barrier of the source junction is reduced, and more electrons are injected into the channel, thus increasing Q_e. This effect is known as drain-induced barrier lowering.

5.7.3 Reduction of V_{th} due to short channel

Figure 5.17 shows the channel region with the source and drain junctions, depletion layers, and the depletion layer corresponding to Q_{Bth} at a voltage equal to V_{th}. Since the potential $2\phi_F$ is close to the built-in barrier of the N^+P source and drain junctions, these three depletion-layer thicknesses are approximately the same and equal to d_{po} for $V_{DB} = V_{SB} = 0$. The amount of channel charge Q'_B over the length L' controlled by the gate bias is less than the previously assumed value $Q_{Bth} = qN_Ad_{po}$ over the length $L'' = L_{mask} - 2r_j$. By assuming that only the charge in the trapezoidal region is controlled by the gate voltage, it can be shown (after Yau [10]) that the threshold voltage is reduced compared to the value given by Eq. (5.7) by an amount ΔV_{thL}, where

$$\Delta V_{thL} = (Q_{Bth}/C_{ox})(r_j/L)[(1 + 2d_{po}/r_j)^{1/2} - 1] \tag{5.56}$$

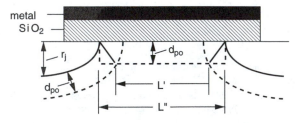

Figure 5.17 Details of channel region for determination of threshold voltage reduction for short channels. $L'' - L_{mask} - 2r_j$.

where r_j is the junction depth, assumed equal to the radius of curvature of the junctions. This tends to zero as d_{po} tends to zero. For a given value of d_{po} and r_j, the magnitude of ΔV_{th} increases for small values of gate length L''.

For devices where the channel width Z is small, a similar effect occurs, but in this case the threshold voltage is increased by an amount (see Neamen [9]):

$$\Delta V_{thZ} = (qN_A d_{po}/C_{ox})(\pi d_{po}/Z) \tag{5.57}$$

This result is strictly valid for the case where the sideways spread of charge Q_B is cylindrical.

5.7.4 Short-channel velocity saturation

The preceding Shockley theory assumes constant carrier mobility in the inversion layer. According to Chapter 2, Fig. 2.10, the velocity starts to saturate at electric fields above 10^4 V/cm. The average value of horizontal field E_{yav} may be estimated as L/V_{DS}, where V_{DS} is close to the pinch-off value for a given V_{GS}. For $L = 1\,\mu m$ and V_{DS} greater than 1 V this will be the case.

In the limit, the current I_{DS} will be given by Eq. (5.21), with the product $\mu_e E_y$ replaced by the saturation velocity v_s. This short-channel value $I_{DS(SC)}$ is thus:

$$I_{DS(SC)} = Zv_s Q_e \tag{5.58}$$

where $Q_e = C'_{ox}(V_{GS}-V_{th})$. In other words:

$$I_{DS(SC)} = Zv_s C'_{ox}(V_{GS}-V_{th}) \tag{5.59}$$

It follows that in this case the current is a linear function of V_{GS}, as shown in Fig. 5.18.

Furthermore, the small-signal transconductance g_m is now given by:

$$g_{m(SC)} = \frac{dI_{DS}}{dV_{GS}} = Zv_s C'_{ox} \tag{5.60}$$

This is constant, in contrast to the Shockley theory g_m for the long-channel MOSFET given by Eq. (5.28), which is proportional to $V_{GS}-V_{th}$.

These conditions may not be completely attained in practice, but in a typical advanced technology submicrometer MOSFET, operation will be closer to the short-channel theory than to the Shockley theory.

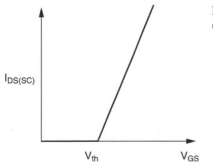

Figure 5.18 $I_{DS(SC)}$ versus V_{GS} for a short-channel (limit-velocity) MOSFET.

The source–drain transit time given by Eq. (5.44) will also be modified for the short-channel device. Since the carriers are now assumed to be traveling at a velocity v_s throughout the channel length L, it follows that the delay time is now:

$$t_{SD(SC)} = L/v_s \qquad (5.61)$$

Example of a short-channel MOSFET:

Consider a MOSFET with a channel length $L = 0.5\,\mu m$, $Z = 10\,\mu m$, $t_{ox} = 50$ nm. With $V_{DS} = 2$ V, this gives an average horizontal field $E_{yav} = 2/0.5 \times 10^{-4} = 4 \times 10^4$ V/cm. This exceeds the "critical" value E_C and the velocity from Fig. 2.10 will be close to 10^7 cm/s.

$$C'_{ox} = 0.3 \times 10^{-12}/50 \times 10^{-7} = 6.0 \times 10^{-8} \text{ F/cm}^2$$

The I_{DS} characteristic is given from Eq. (5.59) by:

$I_{DS} = 10 \times 10^{-4} \times 10^7 \times 6.0 \times 10^{-8}(V_{GS} - 2)$, or $I_{DS} = 0.6 \times (V_{GS} - 2)$ mA.

The transconductance from Eq. (5.60) is: $g_m = 0.6$ mA/V.

The source–drain transit time from Eq. (5.61) is $0.5 \times 10^{-4}/10^7 = 5$ ps. This corresponds to a unity current gain frequency [Eq. (5.43)] $f_t = 1/2\pi t_{sd} = 32$ GHz.

5.7.5 Lightly doped drain (LDD) MOSFET

Examination of Eq. (5.33) indicates that for a given V_{DS}, as w_{scl}/L increases, the output conductance g_{out} increases (the slope of the I_{DS}–V_{DS} characteristic increases). This means that for short-channel lengths, the $I_{DS} = f(V_{DS})$ characteristics will tend not to saturate, as seen in Fig. 5.9 for a short-channel MOSFET. This undesirable effect can be reduced by using the lightly doped drain structure shown in Fig. 5.19.

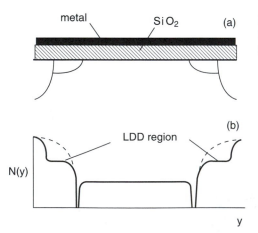

Figure 5.19 Lightly doped drain (LDD) MOSFET: (a) cross section; (b) impurity profile along silicon surface.

In the LDD structure the goal is to reduce the extension of the drain depletion layer into the channel region. According to Eq. (2.56), the depletion layer thickness for a given voltage will be reduced on the lightly doped side if the heavily doped side is less heavily doped. For example from Eq. (2.59) if N_D on the N side is equal to N_A on the P side, the voltage will be equally split between the two sides of the PN junction (instead of almost all being on the lightly doped side). In Fig. 5.19 the conventional single diffused/implanted drain impurity profile in the lateral direction at the surface, shown as a broken line, is replaced by a more lightly doped region, this reducing the depletion-layer penetration into the channel region. This has two desirable effects: First the channel length modulation effect is reduced; w_{scl} in Eq. (5.35) is less, and the output conductance is decreased (V_A is increased); second, the drain–substrate breakdown voltage is increased.

5.8 OVERVIEW OF THE MOSFET CAD SPICE MODEL

Several SPICE models for the MOSFET have been implemented and are available in most SPICE software, unlike the BJT, for which the basic Gummel–Poon model has been used for many years, with some variations and enhancements in particular versions. In the following we will introduce only the most basic MOSFET SPICE parameters. From the earliest versions, it was customary to include the values of Z and L as input parameters, since these determine directly the dc characteristics (and also the f_t) and are available at the mask layout stage of design. This is not the case for the BJT.

The most significant parameters of the MOSFET model used in SPICE are as follows.

- KP is the product $\mu C'_{ox}$ used in Eq. (5.25);
- W is the channel width [Z in Eq. (5.25)] and L is the total channel length;
- TOX is the gate oxide thickness t_{ox}.
- The threshold voltage at zero bias is VT0.

The substrate doping NSUB is the N_A value used in Eq. (5.5) to compute ϕ_F and to obtain the threshold voltage variation with bias contained in Eqs. (5.7) and (5.11).

With these basic parameters, the SPICE model is capable of predicting the main dc characteristics of the MOSFET. In addition, for high-frequency modeling, various capacitance terms must be included. The source and drain diffusions exhibit classical depletion-layer capacitance variation with bias [as in the BJT equations (4.58), (4.59)] and are modeled by their zero-bias capacitances for the area (CJ) and perimeter (CJSW) components. For the source:

$$C_{BS} = CJ \cdot A_S / (1 - V_{BS}/PB)^{MJSW} + CJSW \cdot P_S / (1 - V_{BS}/PB)^{MJSW}$$

For the drain:

$$C_{BDS} = CJ \cdot A_D / (1 - V_{BS}/PB)^{MJSW} + CJSW \cdot P_D / (1 - V_{BS}/PB)^{MJSW}$$

where A_S, A_D are the source and drain areas, PS, PD are the corresponding perimeter lengths. CJ is in F/cm², and CJSW is in F/cm.

Short-channel effects including variation of threshold voltage with V_{DS}, nonconstant I_{DS} with increasing V_{DS} (V_A parameter where the SPICE λ parameter is approximately $1/V_A$) and velocity-field dependence are included with various additional model parameters.

An excellent presentation of both the MOSFET and BJT SPICE models may be found in Ref. 7.

5.9 MOSFET AND BJT COMPARISON

Transconductance

For the BJT the transconductance (4.49) is given by $g_m = I_C/V_t$. Since I_C is an exponentially increasing function of forward V_{BE} bias, it is clear that extremely large values of g_m can be obtained, even for small-area devices. The ultimate limit is set by the high-current effects discussed in Chapter 4. In practice, even for a small VLSI BJT with an emitter area $1 \times 1\ \mu m^2$, I_C can be of order 1 mA, thus making g_m values of 40 mA/V readily attainable.

For the MOSFET, g_m is strictly limited by the geometry as per Eq. (5.28). For $Z = L = 1\ \mu m$ and $t_{ox} = 0.1\ \mu m$, at $V_{GS} - V_{th} = 5$ V, we have $g_m = 0.075$ mA/V.

The low g_m obtainable with MOSFETs is not in itself bad, but for the fact that in order to obtain a reasonable voltage gain ($g_m R_L$) high values of load resistance are necessary; this will result in large RC time constants due to parasitic capacitances.

Switching speed and bandwidth

The upper frequency limit of the BJT is set partly by its f_t, which as given by Eqs. (4.75)–(4.78), is limited by vertical delay times. In particular, the vertical base transit time $t_{bb} = W_b^2/2D_n$ and W_b is readily reduced well below 0.1 μm simply by using shallow junction depths. For the MOSFET, as we saw above, the ultimate speed or bandwidth is related to the f_t and is determined by the source–drain transit time plus significant parasitics. It is inherently more difficult to reduce a lateral device dimension (since it is related to the mask layout and lithography limitations) than to reduce a vertical thickness (dependent on thermal processing). So ultimately, it is easier to obtain high-speed operation from BJTs than from MOSFETs. This BJT advantage is greatly enhanced by the ability to have a high transconductance with the BJT, thus making it easy to drive high capacitive loads.

One might then ask "why is CMOS so widespread?" The answer is partly related to layout considerations. In Chapter 8 we explain that bipolar integrated circuits involve isolation using a deep P^+ diffusion. In MOSFET circuits (including CMOS) this is not necessary, since two MOSFET devices may be placed side by side with very little electrical coupling occurring (in CMOS the problem is considerably complicated by the need for N wells and P wells for the P- and N-channel devices and the need for extra N^+ or P^+ diffusions or oxide isolation to reduce latchup). The problem with bipolar chips is partly also that the high speed is accompanied by high currents and thus by high power dissipation. This can be a severe problem in high-speed ECL. CMOS logic inverters possess the distinct advantage of zero standby power consumption (although this is not as advantageous as might appear at first sight, since in a computer a clocked CMOS chip dissipates more power at higher clock speeds).

A combination of CMOS and bipolar technology called BiCMOS and described briefly in Chapter 8 combines the advantages of both devices.

5.10 CONCLUSIONS

This chapter started with a study of the metal oxide semiconductor (MOS) system, including the MOS capacitance and its variation with bias. The MOSFET structure was then introduced and the static current–voltage relations derived using the Shockley gradual channel, constant-mobility approximation. The small-signal parameters were derived (including output conductance due to channel length modulation) and the small-signal equivalent circuit introduced. After a discussion of MOSFET switching speed, the CMOS inverter was explained and its switching performance discussed briefly. Since in advanced small-geometry sub-micron VLSI MOSFETs second-order effects become important, a section devoted to these effects was presented. Included were the following: inversion layer versus gate voltage and sub-threshold operation; threshold voltage modification due to short-channel effects; short-channel velocity saturation effects. This latter point is very important in small-geometry devices, since it affects both transconductance and delay time. The low doped drain MOSFET was briefly discussed. This was followed by a short comparison of BJTs and MOSFETs, focussing on the essential difference of critical vertical dimensions for the BJT versus critical horizontal dimensions for the MOSFET. The chapter concluded with an introduction to MOSFET SPICE parameters.

PROBLEMS

1. Consider an N-channel aluminum gate MOSFET with the following known quantities: $N_A = 10^{15}$ cm^{-3}, $Q_{ox} = 2 \times 10^{11}$ cm^{-2} (positive charge); $t_{ox} = 0.1\,\mu$m. Calculate the value of threshold voltage V_{th} for $V_s = 0$.

2. Repeat the calculation of Problem 1 for a P-channel device with the same parameter values.

3. Calculate the threshold voltages in Problems 1 and 2 for surface voltages V_s of 1, 2, 4 V.

4. For the MOS structure of Problem 1, calculate the depletion-layer thickness d_p and charge Q_B for $V_S = 0$ and $2\phi_F$. Calculate the corresponding values of V_{GS}.

5. For the MOS structure of Problem 1, calculate the values of C_{ox}, and C_D for $V_S = 0$ and $V_S = 2\phi_F$. Calculate also the total capacitance per cm^2 for large V_{GB} in each case for (a) low frequency; (b) high frequency.

6. An N-channel MOSFET has $V_{th} = 2$ V and $t_{ox} = 0.1\,\mu$m. Assuming a surface mobility value $\mu_n = 500$ cm^2/V s, calculate the drain–source current for a device with $Z = 10\,\mu$m, $L = 10\,\mu$m (minimum design rule) at the following bias values: $V_{GS} = 6$ V, $V_{GS} = 4$ V, at $V_{DS} = 1, 2, 6$ V. Sketch the $I_{DS} = f(V_{DS}, V_{GS})$ (with the pinchoff locus superimposed) and the $I_{DS} = f(V_{GS})$ characteristics.

7. For the MOSFET of Problem 6 at $V_{GS} = 4$ V, at $V_{DS} = 6$ V calculate the values of the small-signal parameters: g_m, t_{sd}, f_t. If $N_A = 10^{15}$ cm^{-3} and the source and drain diffusions are each $10\,\mu$m wide (and $10\,\mu$m in the z direction), estimate the values of source and drain junction capacitances and the input gate capacitance.

8. If the source-to-drain diffusion mask spacing is $L_{mask} = 1\,\mu$m and the junction depths are each 0.1 μm (assume lateral diffusion equal to vertical diffusion) and the channel doping level is $N_A = 10^{17}$ cm^{-3}, calculate the actual channel length L for $V_{DS} = 1$ and 5 V.

9. For the MOSFET of Problem 7, estimate the thickness of the drain–substrate depletion layer at $V_{DS} = 0$ and 6 V using Eq. (3.46). Hence calculate the small-signal output conductance and the MOSFET "Early voltage," V_A.

10. The N-channel device of Problem 7 is to be used in a CMOS inverter. Assuming that a P-channel device with the same oxide thickness, a surface mobility of $\mu_p = 200$ cm^2/V s and an identical threshold voltage is available, calculate the dimensions of the P-channel device for symmetric transfer characteristics. Estimate the f_t of both the N-channel and P-channel devices. Hence estimate the turn-on delays of separate NMOS and PMOS inverters loaded with identical stages. Neglecting junction capacitances, estimate the switching delay of the CMOS inverter with a fan-out of 2 (i.e., loaded by two identical stages).

11. For the CMOS inverter of Problem 10, calculate the two junction capacitances and hence estimate a revised value for the switching times.

12. Consider the design of a CMOS inverter in which the N- and P-channel devices each have a channel doping level of 10^{16} cm^{-3}, each with $t_{ox} = 50$ nm and an interface charge density $Q_{ox} = 10^{11}$ cm^{-2}. Calculate the separate threshold voltages and determine the ion implant dose required for one device to make the threshold voltages equal.

13. If the maximum electric field across the oxide layer limited by dielectric breakdown is 10^6 V/cm, calculate the maximum value of V_{GS} and hence the maximum transconductance for the MOSFET of Problem 6.

14. Use the diode breakdown voltage formula (3.41) to estimate the maximum value of V_{DS} that can be used in the MOSFET of Problem 7.

15. Calculate the nominal V_{th} and short-channel reduced threshold voltages for a MOSFET in which $N_A = 10^{16}$ cm^{-3}, $t_{ox} = 50$ nm, junction depth $r_j = 0.2\,\mu$m, for three values of L: 5, 1, 0.5 μm.

16. Calculate the average value of horizontal electric field for an N-channel MOSFET with $L = 3$, 1.0, 0.3, 0.1 μm assuming V_{DS} is biased at a pinch-off locus value of 4 V. Hence estimate the average electron velocity in each case and decide which theory should be applied to determine the electrical characteristics. If $t_{ox} = 0.1\,\mu$m, $Z = 10\,\mu$m, $V_{GS} = 2$ V, $V_{th} = 1$ V, estimate the values of I_{DS}, g_m, t_{sd}, f_t.

17. Consider a MOSFET with $t_{ox} = 50$ nm, $V_{th} = 1$ V, $V_{GS} = 2$ V, $V_{DS} = 1$ V, $N_A = 10^{17}$ cm^{-3}, $L = 1.0\,\mu$m, $Z = 10\,\mu$m. Calculate I_{DS} and then use Eqs. (5.17) and (5.22) to calculate the magnitude of the inversion-layer charge $Q_e(y)$ and horizontal electric field $E(y)$ at: (a) $y = 0$, $V(y) = 0$; (b) at the value of y corresponding to $V_{DS}/2$; (c) at $y = L$. Comment on the validity of the values thus obtained and estimate more accurate values.

18. Use Fig. 5.16 to estimate values of surface potential V_S and inversion-layer charge Q_e for V_{GB} 2.5, 3.0, 3.5 V. Estimate the electron concentration at the surface for these three conditions and hence in each case calculate the approximate value of surface-layer resistivity, and use the results of Appendix F to compare the surface Fermi level positions.

19. Consider a MOSFET on a substrate doped $N_A = 10^{17}$ cm^{-3} with $L = 0.5\,\mu$m. If $V_{DS} = 3$ V, calculate the value of drain depletion penetration w_{scl} into the channel for (a) a heavily doped drain; (b) an LDD structure with a low-doped region of $N_D = 10^{16}$ cm^{-3} (assume the same value of L in both cases and a junction depth of 0.1 μm). Calculate also the output resistance in each case for $I_{DS} = 100\,\mu$A using Eq. (5.33).

References

1. E. S. Yang, *Microelectronic Devices*. New York: McGraw–Hill, 1988.
2. B. G. Streetman, *Solid State Electronic Devices*. Englewood Cliffs, New Jersey: Prentice-Hall, 1990.
3. A. Bar-Lev, *Semiconductors and Electronic Devices*, 3rd Ed. New York: Prentice-Hall, 1993.
4. D. H. Navon, *Semiconductor Microdevices and Materials*. New York: Holt, Rinehart, Winston, 1986.
5. M. S. Tyagi, *Introduction to Semiconductor Materials and Devices*. New York: Wiley, 1991.
6. D. L. Pulfrey. N. G. Tarr, *Introduction to Microelectronic Devices*. Englewood Cliffs, New Jersey: Prentice-Hall, 1989.
7. P. Antognetti, G. Massobrio, *Semiconductor Device Modeling with SPICE*. New York: McGraw–Hill, 1987.
8. Y. P. Tsividis, *Operation and Modeling of the MOS Transistor*. New York: McGraw–Hill, 1987.
9. D. A. Neamen, *Semiconductor Physics and Devices*. Chicago: Irwin, 1997.
10. L. D. Yau, "A simple theory to predict the threshold voltage of short channel IGFETs," *Solid State Electronics* **17**, 1059 (1974).

Chapter 6

Junction Field-Effect Transistors

6.1 INTRODUCTION

The junction field-effect transistor (JFET) and the MOSFET have in common the fact that an electric field perpendicular to the semiconductor surface is used to control current flow in the horizontal direction. In the JFET, however, the field occurs across a conventional PN junction depletion layer in the bulk region, and there is no gate oxide layer. The basic behavior is similar in both devices, and the current flows by majority carriers only, with a pinch-off region near the drain side of the conducting channel. The technology is, however, quite different for the two devices, and the theoretical derivations are not identical. In this chapter we derive the static current–voltage characteristics using the Shockley gradual channel approximation. The equivalent circuit is explained and high-frequency behavior analyzed, leading to the f_t, as for the MOSFET. The nature of pinch-off is discussed in some detail because it is important in microwave structures. This leads to discussion of the limit velocity theory, similar to the short-channel MOSFET. The metal semiconductor FET, or MESFET, is almost identical to the JFET, and no additional theory is required; however, a brief overview of the MESFET device is included.

6.2 JFET STRUCTURE

We reproduce below the cross section of the JFET used in Chapter 1. The device is assumed to be made by growing a thin N-type epitaxial layer on a lightly doped P^- substrate. The two N^+ diffusions provide ohmic contacts to the source and drain terminals. The P diffusion creates a P–N junction, with the P region being contacted to the gate terminal. The source, gate, and drain diffusions and contacts extend parallel to each other in the z direction (Fig. 6.1). Under normal operation a positive voltage V_{DS} is applied between drain and source and a reverse (negative) bias V_{GS} between gate and source, as shown in Fig. 6.2.

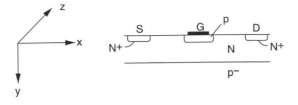

Figure 6.1 Cross section of an *N*-channel JFET on a P^- substrate. The three terminals are: source, gate, drain.

Figure 6.2 Simple *N*-channel JFET bias arrangement.

6.3 dc CHARACTERISTICS OF THE JFET

6.3.1 Low drain–source bias

Let us examine first of all what happens with only a very small V_{DS} bias applied. Figure 6.3 shows four depletion-layer boundaries as the reverse bias V_{GS} goes from zero to a reverse bias of a few volts. At zero bias, the vertical depletion-layer thickness extending downwards into the channel is given by the one-sided junction approximation (2.62), with the total voltage equal to the built-in barrier potential V_{bi}:

$$d_n = \sqrt{(2\epsilon V_{bi}/(qN_D)} \tag{6.1}$$

As the reverse voltage increases, a value V_p is reached at which the depletion-layer extends all the way across the channel thickness a. From the same equation (2.62) we obtain:

$$V_p + V_{bi} = (q/2\epsilon)N_D a^2 \tag{6.2}$$

Throughout the following discussion it will be assumed that the P^- substrate is very lightly doped so that the depletion-layer of the bottom junction extends upwards by a negligible amount, as given by Eqs. (2.58) and (2.59).

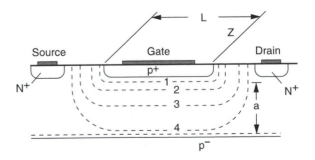

Figure 6.3 Depletion-layer boundaries from 1 for zero V_{GS} to 4 for large reverse V_{GS} bias.

Let us now consider the channel resistance R_C seen between source and drain. This is given by:

$$R_C = V_{DS}/I_{DS} \tag{6.3}$$

The value of R_C is determined by the channel length L (here we neglect the small resistances on the source and drain sides of the channel), width Z, the resistivity $\rho_c = 1/(q\mu_n N_D)$, and the vertical thickness of the conducting part of the channel, which is simply $a - d_n$. We thus obtain:

$$R_C = [L/(q\mu_n N_D)]/[(a - d_n)Z] \tag{6.4}$$

where Z is the length of the device perpendicular to the cross section in Fig. 6.3. Clearly, this resistance has its lowest value for zero applied bias and reaches an infinite value for $V_{GS} = -V_p$. Figure 6.4 illustrates the corresponding I_{DS} versus V_{DS} curves, where V_{DS} values much less than V_p are considered. The lowest value of $R_C = R_{CO}$ occurs for the zero-bias case (assuming $d_n << a$ at zero bias):

$$R_{CO} = L/(q\mu_n N_D a Z) \tag{6.5}$$

Curve 4 is shown for infinite resistance, that is, when $d_n = a$. For JFET structures this is referred to as the channel pinchoff condition, and the corresponding gate–source voltage is called the pinch-off voltage, $V_{gp}(= -V_p)$. Under this bias condition, the conducting region of the channel has been pinched off; so no current can flow (or rather, only a small leakage current flows).

6.3.2 Large drain–source bias

Let us now consider what happens when the voltage between drain and source V_{DS} increases in the positive direction. Note that a positive voltage on the drain side of the gate implies a reverse bias between the drain side of the conducting channel and the P diffusion of the gate. Initially, we consider only the case of zero V_{GS} bias. Figure 6.5 shows the depletion-layer

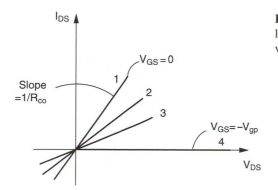

Figure 6.4 I_{DS} versus V_{DS} characteristics for low V_{DS}, showing how the channel resistance varies with reverse V_{GS} bias.

boundaries of the *P–N* junction as V_{DS} is increased. Curve A is the same as curve 1 of Fig. 6.3. For a moderate V_{DS} value (typically less than 1 V), the voltage drop in the channel due to R_C must also appear as a vertical voltage across the depletion-layer (we keep $V_{GS} = 0$). This forces a reverse bias across the depletion-layer, making it wider at the drain side ($x = L$) of the channel. We know that if a voltage V_p is applied across the depletion-layer, it will extend across the channel thickness a. At some (as yet unknown) current I_{DS}, case D is thus reached, where the conducting channel is just pinched-off at the drain end; this occurs at the drain–source pinch-off voltage $V_{DS} = V_{DSP}$. In this case we have $V_{DSP} = V_p$.

It is clear that the drain–source resistance R_C of the conducting region of the channel is higher as V_{DS} increases, since the average vertical thickness $(a - d_n)$ of the conducting channel decreases from left to right in Fig. 6.5. The current I_{DS} therefore rises less rapidly with V_{DS} as V_{DS} increases. Figure 6.6 illustrates this behavior. Points A through D correspond to the same labels for depletion-layer boundaries in Fig. 6.5. The gradient at low V_{DS} bias is clearly $1/R_{CO}$, with R_{CO} given by Eq. (6.5). Since the right-hand side of the channel is pinched off at $V_{DS} = V_{DSP}$, the current can no longer increase and is said to be saturated. In fact, it is clear that complete pinch-off cannot occur. What actually happens is that as the conducting region narrows, the horizontal electric field increases (to maintain a constant current); the electron velocity increases, eventually saturating as explained in Section 2.3.1, Fig. 2.10. Under certain conditions (yet to be determined) it is a fair approximation to assume that the thickness of the conducting channel $\delta = a - d_n$ is much less than a at the drain end of the channel; so for the moment we set $\delta = 0$.

The question to be answered is "What is the maximum value of I_{DS}, frequently referred to as I_{DSS}?" Clearly this maximum value must be less than V_p/R_{CO}, as shown in Fig. 6.7, since

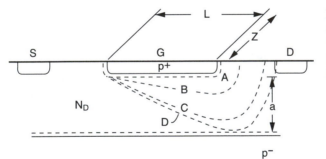

Figure 6.5 Variation of gate-to-channel depletion-layer boundary as V_{DS} is increased from zero (broken line A) to V_{DSP} (broken line D), keeping $V_{GS} = 0$.

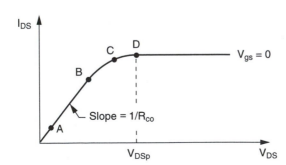

Figure 6.6 I_{DS} versus V_{DS} curve for $V_{GS} = 0$. Labels A through D correspond to the same labels for the depletion-layer boundaries as shown in Fig. 6.5.

R_C increases with increasing V_{DS}. An estimate of the shape of the curve can be obtained by considering a voltage $V_{DS} = V_{DSP}/2$. The depletion-layer thickness at the drain side ($x = L$) is then $y = a/\sqrt{2} = 0.707a$, and the conducting channel thickness at $x = L$ is $0.29a$. The average channel thickness will be between $0.29a$ and a, and the resistance R_C will hence be of order $2R_{CO}$; the I_{DS} versus V_{DS} curve will be visibly below the straight line of slope $1/R_{CO}$. In fact, as we will now proceed to show, the maximum value of current is $I_{DSS} = V_{DSP}/3R_{CO}$.

6.3.3 The Shockley theory for JFET *I–V* characteristics

Figure 6.8 shows the depletion-layer (shaded) region of vertical width y, for an arbitrary V_{GS}, V_{DS} bias. dx is an increment in the horizontal x direction between $x = 0$ and L. We assume the following:

1. The conducting channel has a mobility that is constant from $x = 0$ to L;
2. The conducting channel becomes completely pinched off at $x = L$ for $V_{DS} + V_{SG} = V_p$.
3. The conducting channel is a neutral region ($dE/dx = 0$).

These conditions are valid for what is referred to as the *gradual channel approximation*.

The resistance corresponding to an incremental distance dx, where the channel thickness is $a - y$, is given by:

$$dR = dx/(a - y)\sigma Z \tag{6.6}$$

where $\sigma = q\mu_n N_D$. The voltage drop $\int I_{DS}\, dR$ in the channel from 0 to x is thus:

$$V_x = \frac{I_{ds}}{\sigma Z} \int \frac{dx}{a - y} \tag{6.7}$$

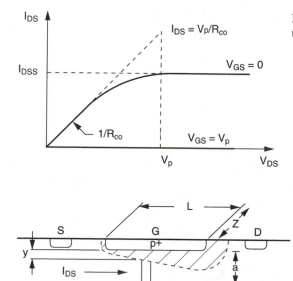

Figure 6.7 I_{DS} versus V_{DS} for estimation of the saturation current I_{DSS}.

Figure 6.8 Diagram used in Shockley derivation of JFET *I–V* characteristics.

But from Kirchhoff's voltage law, summing the potential from the source, along the channel, upwards across the depletion-layer and back to the source, we have:

$$V_x + (V_{SG} + V_{bi}) = V_y = (q/2\epsilon)N_D y^2 \tag{6.8}$$

Hence we obtain:

$$V_x = I_{ds}/(\sigma Z) \int [1/(a - y)]\, dx = (q/2\epsilon)N_D y^2 - (V_{SG} + V_{bi}) \tag{6.9}$$

Integrating this equation from $x = 0$ to L corresponding to the voltage varying from $V_x = 0$ to V_{ds} yields the result:

$$I_{DS} = \frac{1}{R_{CO}}\left[1 - \frac{2V_p}{3V_{DS}}\left(\frac{V_{SG} + V_{bi} + V_{DS}}{V_p}\right)^{3/2} + \frac{2V_p}{3V_{DS}}\left(\frac{V_{SG} + V_{bi}}{V_p}\right)^{3/2}\right] V_{DS} \tag{6.10}$$

This equation describes the I_{DS}–V_{DS} characteristics as shown in Fig. 6.9. Note that the equation is only valid up to the pinch-off point, where the gradient becomes zero. Setting $dI_{DS}/dV_{DS} = 0$ in Eq. (6.10) and neglecting the V_{bi} terms for simplicity gives:

$$0 = (1/R_{CO})\{1 - [(V_{SG} + V_{DS})/V_p]^{1/2}\} \tag{6.11}$$

In other words the "pinch-off" locus is given by:

$$V_{SG} + V_{DS} = V_p \tag{6.12}$$

A simplified equation may be obtained for the top curve corresponding to $V_{SG} = 0$. In this case we obtain the result:

$$I_{DS}/V_{DS} = (1/R_{CO})[1 - (2/3)(V_{DS}/V_p)^{1/2}] \tag{6.13}$$

A final very important result is the transconductance $g_m = dI_{DS}/dV_{GS}$. This may be obtained for any point on or beyond (i.e., to the right of) the pinchoff locus by calculating g_m for the pinch-off condition given by Eq. (6.12). Substituting Eq. (6.12) into Eq. (6.10) gives:

$$I_{DS(pinchoff)} = (V_p/3R_{CO})[1 - 3(V_{SG}/V_p) + 2(V_{SG}/V_p)^{3/2}] \tag{6.14}$$

Differentiating to obtain the transconductance gives:

$$g_m = \frac{dI_{DS(pinchoff)}}{dV_{SG}} = (1/R_{CO})[(V_{SG}/V_p)^{1/2} - 1] \tag{6.15}$$

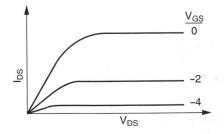

Figure 6.9 I_{DS}–V_{DS} characteristics of a JFET obtained from the Shockley theory.

The maximum possible value of transconductance is obtained when $V_{SG} = 0$. The magnitude of this maximum value is seen from Eq. (6.15) to be:

$$g_{m\,max} = 1/R_{CO} \qquad (6.16)$$

This result is characteristic of FET structures—as in the MOSFET, the maximum transconductance is the slope of the I_{DS}–V_{DS} characteristic at low V_{DS}. Since R_{CO} is a function of doping level and the L and Z dimensions of the structure, and since the maximum value of doping level is limited by the desired breakdown voltage (as in the collector of a BJT), it is clear that as in the MOSFET, the scope for obtaining large g_m is very limited. This result confirms one of the major advantages of a BJT compared to a FET device, the fact that in the BJT, the transconductance increases exponentially with V_{BE}, as indicated by Eq. (4.48) and can thus be much greater than the g_m of an equal-area JFET.

We also note that the pinch-off voltage V_p in the above equations is analogous to the threshold voltage V_{th} in the MOSFET. Although the depletion JFET discussed above is the most common structure, it is also possible to have a normally "off" or enhancement JFET in which at zero bias the depletion-layer extends across the channel, thus making I_{DS} zero. Note, however, that only low forward bias (of order 0.5 V) may be applied before the gate current starts increasing exponentially with voltage according to the normal diode law.

6.4 THE DOUBLE-GATE JFET

If the JFET of Fig. 6.1 is fabricated on a substrate with a high doping level, then both gates may be used as input terminals, or G_2 grounded. The depletion-layers extend from each junction, as shown in Fig. 6.10 (where negligible penetration is assumed in the heavily doped P regions).

Using the same definition of channel thickness a as previously, the gate pinch-off voltage applied simultaneously from the source to G_1 and G_2 to cause both depletion-layers to meet at a vertical thickness $a/2$ will be from Eq. (6.2):

$$V_p + V_{bi} = (q/2\epsilon)N_D(a/2)^2 \qquad (6.17)$$

This is smaller by a factor 1/4 than the previous value for a single-gate structure. It can be shown that the saturation current I_{DSS} is also smaller by the same factor. The maximum

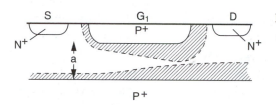

Figure 6.10 Double-gate JFET structure.

transconductance remains the same $(1/R_{CO})$, as given by Eq. (6.16). If used in the parallel connection with a heavily doped P^+ substrate, the input capacitance will be more than double that for the single-gate connection (because of the larger area of the channel to substrate junction). However, it is sometimes useful in circuits to have the fourth terminal available so that separate input signals may be applied.

6.5 SMALL-SIGNAL EQUIVALENT CIRCUIT AND HIGH-FREQUENCY PROPERTIES

The small-signal equivalent circuit of the JFET can be drawn as in Fig. 6.11, where C_{gs} is the gate–source capacitance determined by the depletion-layer. r_{gs} is the small-signal value of the channel resistance. C_{gd} is the capacitance associated with the depletion-layer on the drain side of the pinch-off point to the right of the gate. Apart from g_m, the values of the small-signal parameters can only be estimated.

The value of C_{gs} may be taken from the "average" value of the depletion-layer thickness under the gate. Assuming that pinch-off has occurred, this thickness will be of order $a/2$ for normal bias in the saturation region, and the corresponding capacitance may be written:

$$C_{gs} \approx 2\epsilon L Z/a \tag{6.18}$$

The value of C_{gd} may be estimated from the drain sidewall component of the P–N junction depletion-layer, as shown in Fig. 6.8. If V_{DS} is such that pinch-off has just occurred, the depletion-layer boundary will be roughly that of the arc of a quarter circle. The "area" of the capacitance is $\pi a Z/2$, and the "thickness" of the depletion-layer is a. The value of C_{gd} is thus:

$$C_{gd} \approx \epsilon \pi a Z/2a \approx \epsilon \pi Z/2 \tag{6.19}$$

This gives a ratio $C_{gd}/C_{gs} = \pi a/4L$, which is roughly equal to a/L and usually considerably less than unity.

It should be noted that both these capacitances are really bias dependent. C_{gs} will decrease below the given value for large reverse V_{GS} bias, but its minimum value based on the depletion-layer thickness cannot be less than $\epsilon L Z/a$. The value of C_{gd} will decrease below the above value as V_{DS} increases beyond the pinchoff value; one would expect an approximate inverse square root dependence on total gate–drain voltage.

The value of r_{gs} may be estimated by referring to the distributed RC system shown in Fig. 6.12. An initial value would be to set $r_{gs} = R_{CO}$. However, the actual value of dc resistance in the wedge-shaped region will clearly be greater than R_{CO}. On the other hand, the rf signal path is such that not all the resistance will be traversed by the total gate current, and the resistance is thus decreased. The result $r_{gs} = R_{CO}$ is therefore a fairly good approximation.

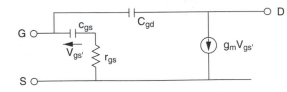

Figure 6.11 Small-signal equivalent circuit of the JFET.

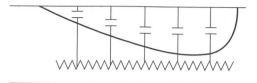

Figure 6.12 Gate–source distributed RC system for estimating the value of r_{gs}. The capacitances are inside the depletion-layer of Fig. 6.5, and the distributed resistance is in the conducting channel.

Using the equivalent circuit of Fig. 6.11, we may now determine (within the limits of the approximations used above) the unity current gain frequency f_t of the JFET.

The output ac current flowing in a short circuit (as in Section 4.6.1 for the BJT analysis) is given by:

$$i_{out} = g_m v'_{gs} \qquad (6.20)$$

where we have used v'_{gs} instead of v_{gs} because the JFET g_m is due to a voltage change across the depletion-layer.

The input current at low or moderate frequencies (such that $\omega C_{gs} r_{gs} \ll 1$) is given by:

$$i_{in} = v'_{gs} \omega C_{gs} \qquad (6.21)$$

The current gain is thus:

$$A_i(\omega) = i_{out}/i_{in} = g_m/\omega C_{gs} \qquad (6.22)$$

The unity current frequency f_t is obtained when $|A_i(\omega)| = 1$, giving

$$f_t = \omega_t/2\pi = g_m/2\pi C_{gs} \qquad (6.23)$$

A good high-frequency JFET will thus have a large g_m and a small C_{gs}. Note that since both g_m and C_{gs} are proportional to the Z length of the device, increasing f_t is obtained primarily by reducing the channel length L.

6.6 SHORT-CHANNEL EFFECTS AND THE MEANING OF PINCH-OFF

Let us now consider in more detail exactly what happens at the drain side of the conducting channel when pinch-off occurs. Figure 6.13 shows a vertical region extending over a distance δ through which all the drain current flows. The magnitude of this current can be expressed in terms of the drift velocity v_d of the electrons:

$$I_{DS} = qZ\delta N_D v_d \qquad (6.24)$$

It is clear that as pinch-off is approached (which we have previously assumed to be when δ becomes zero), the value of velocity must increase. However, we know from our earlier discussion in Section 2.3.1 and Fig. 2.10 that the drift velocity saturates at a value $v_s = 10^7$ cm/s for electrons in silicon. This means that for a given doping level and current, we can determine the value of δ:

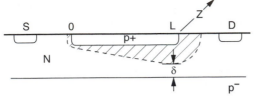

Figure 6.13 Illustration of finite channel thickness at pinch-off.

$$\delta = I_{DS}/(qZN_D v_s) \tag{6.25}$$

It is now clear that only if the residual conducting channel thickness is much less than the zero-bias channel thickness a will the Shockley theory be valid. Substituting the previous result for I_{DS} at the maximum I_{DSS} value gives:

$$\delta = (V_p/3R_{CO})/(qZN_D v_s) \tag{6.26}$$

Substituting from Eq. (6.2) for V_p (neglecting the barrier voltage V_{bi}) and from Eq. (6.5) for R_{CO}, gives:

$$\delta = (\sigma a^3/L)(1/6\epsilon v_s) \tag{6.27}$$

where we have substituted $\sigma = q\mu_n N_D$. It is more instructive to study the ratio δ/a:

$$\delta/a = (\sigma a^2/L)(1/6\epsilon v_s) \tag{6.28}$$

If this ratio is much less than unity, the Shockley theory is valid.

Let us take some typical JFET data: $N_D = 2 \times 10^{15}$ cm^{-3} [which gives $\sigma = 0.4$ (ohm cm)$^{-1}$], $a = 1.4\,\mu$m, $L = 20\,\mu$m.

Using $\epsilon = 10^{-12}$, F/cm, $v_s = 10^7$ cm/s, gives $\delta/a = 0.065$.

This is a fairly good approximation to a completely pinched-off channel.

However, it is clear that if L is decreased to $2\,\mu$m, the ratio δ/a rises to 0.65 and the approximation of a pinched-off channel is simply not valid.

Using Eq. (6.28) it is possible to write the condition for the Shockley theory to be valid:

$$\sigma a^2/L \ll 6\epsilon v_s = 6 \times 10^{-5} \text{ F/s} \tag{6.29}$$

Thus it is not simply the channel length, but this length combined with the channel thickness and conductivity, that determine whether a particular JFET structure obeys the Shockley theory or the short-channel (limit velocity) theory.

Since the trend in JFET design is to use small values of L, clearly it is desirable to be able to estimate the performance for the case of such a "short-channel" JFET.

We may approach this problem by making the opposite assumption to a "pinched-off" channel, that is, assume it is completely open at the drain side. The only mechanism leading to saturation of the drain–source current I_{DS} is now the saturation of the drift velocity at a value v_s. Figure 6.14 illustrates this situation. If the channel thickness a is large enough, or the doping level N_D high enough, saturation of the I_{DS}–V_{DS} characteristic can occur due primarily to velocity-field saturation. In the extreme case where the channel thickness is almost equal to a, we can write the value of maximum drain–source current as:

$$I_{DSS} = qZaN_Dv_s \qquad (6.30)$$

It is important to note that there are now two different definitions for pinch-off voltage. We can define the voltage V_{DSP} at which the I_{DS}–V_{DS} characteristic saturates; this will be different from the gate–source voltage V_{GSP} required to reduce the drain current to zero. The value of V_{DSP} can be estimated from the velocity-field characteristic by taking the critical E_c to define the onset of saturation. In this case we obtain, for $V_{GS} = 0$:

$$V_{DSP} = E_cL \qquad (6.31)$$

The gate–source pinchoff voltage V_{GSP} will be the same as V_p defined by Eq. (6.2).

The dependence of I_{DS} on V_{SG} can be determined by writing the conducting channel thickness as $a - d_n$:

$$I_{DS} = I_{DSS}(1 - d_n/a) \qquad (6.32)$$

or

$$I_{DS} = I_{DSS}\{1 - [(V_{SG} + V_{bi})/(V_p + V_{bi})]^{1/2}\} \qquad (6.33)$$

By differentiating this we find the transconductance:

$$g_m = (I_{DSS}/2)/[(V_{SG} + V_{bi})(V_{pg} + V_{bi})]^{1/2} \qquad (6.34)$$

The maximum value occurs when $V_{SG} = 0$.

A general charge relation for FETs

As for MOSFETs, a general relation for the transconductance of JFETs may be obtained by writing:

$$g_m = \frac{dI_{DS}}{dV_{GS}} = \frac{dI_{DS}}{dQ_{DS}}\frac{dQ_{DS}}{dV_{GS}} \qquad (6.35)$$

where dQ_{DS} is the incremental change in majority carrier charge in the conducting channel due to an incremental change dV_{GS} in V_{GS}. This incremental charge is shown by the shaded area in Fig. 6.15. It is important to note that an increase in conducting channel charge by an

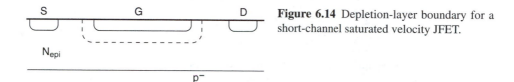

Figure 6.14 Depletion-layer boundary for a short-channel saturated velocity JFET.

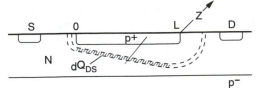

Figure 6.15 Cross section of JFET showing incremental charge $dQ_{GS} = -dQ_{DS}$.

amount dQ_{DS} corresponds to an identical decrease in depletion-layer charge dQ_{GS}. Equation (6.35) can thus be rewritten as:

$$g_m = \frac{dI_{DS}}{dQ_{DS}} \frac{dQ_{GS}}{dV_{GS}} \tag{6.36}$$

We can now identify the two terms on the right-hand side. The first, dQ_{DS}/dI_{DS}, is the channel transit time t_{sd}. The second, dQ_{GS}/dV_{GS}, is the gate–source capacitance C_{gs}. We thus obtain the general relation:

$$g_m = C_{gs}/t_{sd} \tag{6.37}$$

It is interesting to observe by comparison with the expression for f_t (6.23) that the channel transit time and f_t are directly related:

$$f_t = 1/2\pi t_{sd} \tag{6.38}$$

This is analagous to the expression for the MOSFET (5.43) and to the f_t of the BJT (4.75), where we had

$$f_{t(BJT)} = g_m/2\pi C_{\pi t} \tag{6.39}$$

In the case of maximum f_t for the BJT, where diffusion capacitance dominated over junction (depletion-layer) capacitances, as given by Eq. (4.75), we have:

$$f_{t\,max\,(BJT)} = 1/2\pi t_f \tag{6.40}$$

The major practical difference between the apparently similar results for the JFET and the BJT is that t_{sd} in the JFET is a horizontal transit time determined by the gate length L, whereas t_f in the BJT is a vertical transit time determined by the vertical junction depths and collector–base depletion-layer thickness. It is intrinsically easier to reduce vertical junction depths, which are independent of lithography, compared to reducing lateral lengths.

We can estimate the source–drain transit time for the two types of JFET analyzed above. First, for a Shockley theory (long-channel) device where constant field-independent mobility is assumed:

$$t_{sd\,(Shockley)} = L/v_d = L/(\mu_n E_x) \tag{6.41}$$

The horizontal value of electric field can be approximated as $V_{DS}/L = V_p/L$ at pinch-off, giving:

$$t_{\text{sd (Shockley)}} = L^2/(\mu_n V_p) \tag{6.42a}$$

For a short-channel JFET, the velocity is saturated at a value v_s and the transit time is given by:

$$t_{\text{sc (shortchannel)}} = L/v_s \tag{6.42b}$$

As an example, consider the previous JFET structures, $N_D = 2 \times 10^{15}$ cm^{-3}, $a = 1.4\,\mu$m, with $L = 20$ and $2\,\mu$m.

For $L = 20\,\mu$m, where the Shockley theory is valid, the pinch-off voltage from Eq. (6.2) is $V_p = (q/2\epsilon)N_D a^2 = (1.6 \times 10^{-19}/2 \times 10^{-12}) \times 10^{15} \times (2 \times 10^{-4})^2 = 3.2$ V. Using Eq. (6.42a), this gives $t_{\text{sd}} = L^2/\mu_n V_p = (10^{-3})^2/1200 \times 3.2 = 2.6 \times 10^{-10} = 260$ ps.

The corresponding value of f_t is $1/2\pi t_{\text{sd}} = 612$ MHz.

For $L = 2\,\mu$m, the velocity is nearly saturated throughout the channel length L, and Eq. (6.42b) gives:

$$t_{\text{sd}} = L/v_s = 2 \times 10^{-4}/10^7 = 2 \times 10^{-11} = 20 \text{ ps}.$$

The corresponding value of f_t is 8 GHz.

6.7 GaAs MESFET

We introduce this device here because of its widespread use in microwave applications and because in theory and appearance it is almost the same as the JFET. A cross-sectional view is shown in Fig. 6.16.

N-type gallium arsenide is used for the channel because of its very high electron mobility (8500 cm^2/V s for low doping, 5000 cm^2/V s at doping levels of order 10^{17} cm^{-3}). Also the drift velocity peaks at approximately 2×10^7 cm/s (see Fig. 7.27). The constant-mobility Shockley theory predicts a fivefold improvement in frequency response over silicon; the short-channel

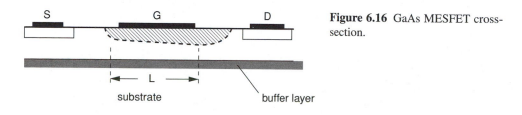

Figure 6.16 GaAs MESFET cross-section.

limit velocity theory predicts up to a twofold improvement. The gate is formed by a metal–semiconductor (Schottky) contact, which reduces the junction capacitance because the junction sidewall components are eliminated; this also allows simpler processing. A semi-insulating GaAs substrate is used with a thin buffer layer (reducing the effects of defects in the substrate). The resulting MESFET has excellent dc and hf characteristics.

Consider a MESFET with a gate length $L = 0.2\,\mu$m doped $N_D = 10^{17}$ cm^{-3} with a channel thickness $a = 0.14\,\mu$m. From Fig. 2.11 the electron mobility $\mu_n = 5000$ cm^2/V s. The channel conductivity $\sigma = q\mu_n N_D = 1.6 \times 10^{-19} \times 5000 \times 10^{17} = 80$ ohm cm^{-1}. Equation (6.29) gives a "crossover condition" for $a = (6 \times 10^{-5} \times 0.2 \times 10^{-4}/80)^{1/2} = 0.37 \times 10^{-5}$ cm. In other words for $a = 0.14\,\mu$m the MESFET should not obey the Shockley theory.

The gate pinch-off voltage from Eq. (6.2) is $V_p = 1.6$ V. For $V_{DS} = V_p = 1.6$ V, the average electric field is $E_x = V_p/L = 1.6/0.2 \times 10^{-4} = 8 \times 10^4$ V/cm. Note that the critical field $E_C = v_s/\mu_n = 2 \times 10^7/5000 = 4000$ V/cm. The average velocity is therefore close to the saturated drift value. Equation (6.42a) for t_{sd} is therefore not valid. From Eq. (6.42b) we find $t_{sd} = L/v_s$ for a limit velocity FET. We can therefore reasonably estimate the actual t_{sd} value to be $t_{sd} = 2 \times 0.2 \times 10^{-4}/2 \times 10^7 = 1 \times 10^{-12}$ s or 1 ps. This translates Eq. (6.38) into an f_t of order 80 GHz.

6.8 CONCLUSIONS

We have seen that the JFET is a simple device to understand, but that the classical Shockley gradual channel theory only applies accurately to long-channel devices satisfying Eq. (6.29) $\sigma a^2/L \ll 6\epsilon v_s = 6 \times 10^{-5}$ F/s. Even in this case, exact simple expressions for the equivalent circuit parameters other than g_m are not available. We examined the short-channel JFET and determined a general relation between input capacitance, transconductance, and transit time. Since JFET technology is closer to bipolar technology than to that for MOSFETs, we can summarize JFET/MESFET performance compared to BJT performance. The main conclusions are that: (1) Their transconductance is inherently limited by geometry, unlike the BJT, where very large g_m values can be obtained for a given geometry; (2) the transit time and thus the f_t of the JFET is determined by a horizontal lithography length, whereas in the vertical *NPN* BJT these quantities are determined by vertical depths unrelated to lithography.

Probably the most important commercial device is the microwave GaAs MESFET used for low-noise amplifiers with useful power gain, although the silicon JFET can be made in integrated form and offers some advantages (very high input impedance at low frequencies) compared to the BJT.

PROBLEMS

1. Derive the expression for I_{DS} given by Eq. (6.10), using Eq. (6.7) to express I_{DS} as a function of V_x and Eq. (6.8) to replace y by its function of V_x.

2. Use Eq. (6.14) to derive the transconductance g_m given by Eq. (6.15).

3. Consider a silicon JFET with an N-layer doped 10^{15} cm^{-3} to a depth of 4.5 μm. Assume a very lightly doped P-type substrate. The gate is formed by a 100 μm (Z) × 40 μm (L) P^+ diffusion to a depth of 0.5 μm. Assume the Shockley theory to be valid and calculate the following quantities: R_{CO}, $g_{m(max)}$, V_{gsp}. Calculate also the ratio δ/a to determine if the FET should obey the Shockley gradual channel theory. Estimate the values of the small signal equivalent circuit parameters r_{gs}, C_{gs}, C_{gd} and calculate f_t.

4. In the above JFET, if the P substrate is doped 10^{14} cm^{-3}, calculate the N-region depletion-layer thickness of the P–N junction at each end of the channel for $V_{SG} = 0$, $V_{DS} = V_p$, and estimate the value of C_{gs}.

5. In the JFET of Problem 3, if the substrate doping is increased to 10^{18} cm^{-3} to make a double-gate structure, calculate approximately the value of total input capacitance at pinch-off ($V_{GS} = 0$) and compare with the value obtained in Problem 4 (assume the total length of the substrate is $3L$).

6. Calculate the parameters of the JFET of Problem 3 if the doping level is 10^{16} cm^{-3}, the total N-layer thickness is 1.5 μm, the channel length $L = 1$ μm. Repeat the calculation of depletion-layer thickness as in Problem 4.

7. Starting with the JFET of Problem 6 with a channel doping level of 10^{16} cm^{-3}, calculate and plot the maximum transconductance (make suitable estimates where necessary) for residual channel thickness a varying from 0.1 to 3 μm using the Shockley or limit velocity theory as appropriate.

8. A GaAs MESFET is designed with a channel length $L = 1$ μm, $Z = 50$ μm, channel thickness $a = 0.5$ μm doped 5×10^{15} cm^{-3}. Use Eq. (6.29) to estimate whether this MESFET should obey the Shockley or short-channel theory. Hence estimate the value of f_t and the maximum small-signal transconductance (use Fig. 2.11 to determine the channel mobility and hence the conductivity).

References

1. E. S. Yang, *Microelectronic Devices*. New York: McGraw–Hill, 1988.
2. B. G. Streetman, *Solid State Electronic Devices*. Englewood Cliffs, New Jersey: Prentice-Hall, 1990.
3. A. Bar-Lev, *Semiconductors and Electronic Devices*, 3rd Ed. New York: Prentice-Hall, 1993.
4. D. H. Navon, *Semiconductor Microdevices and Materials*. New York: Holt, Rinehart, Winston, 1986.
5. M. S. Tyagi, *Introduction to Semiconductor Materials and Devices*. New York: Wiley, 1991.
6. D. L. Pulfrey and N. G. Tarr, *Introduction to Microelectronic Devices*. Englewood Cliffs, New Jersey: Prentice-Hall, 1989.
7. D. A. Neamen, *Semiconductor Physics and Devices*. Chicago: Irwin, 1997.

Chapter 7

Overview of Special-Purpose Semiconductor Devices

7.1 INTRODUCTION

This chapter introduces the reader to a range of "specialist" semiconductor devices. Although the MOSFET and the bipolar transistor are the essential components of VLSI silicon chips, there are a number of other devices that are essential to a wide range of electronic systems. Optoelectronic devices are extremely important today with applications in fiber-optic detectors and transmitters and in imaging arrays; so we commence with these topics. This is followed by sections on power, microwave, and memory devices. Mathematical treatment will be limited, and only an outline of the theory of each device will be provided. The goal is to present the basic concepts of each device in such a manner that the reader can grasp the fundamentals. In keeping with the philosophy of this book, for a more detailed treatment of each device, the reader is referred to the more advanced or more specialist texts.

7.2 OPTOELECTRONIC DEVICES

7.2.1 The photoconductor detector

A bar of semiconductor material of constant doping level as shown in Fig. 7.1 may be used as a photodetector. The photoconductivity of the semiconductor is determined by the excess photogenerated carriers with concentrations n' and p' generated by the photon flux density ϕ (photons per square cm) incident on the area A_ϕ. Assuming uniform carrier generation, the conductivity is given by:

$$\sigma = q[\mu_n(n_0 + n') + \mu_p(p_0 + p')] \tag{7.1}$$

where n_0, p_0 are the thermal equilibrium concentrations and μ_n, μ_p are the electron and hole mobilities. Excess electrons and holes are generated in equal concentrations $n' = p'$, by the incident light, and since $n_0 p_0 = n_i^2$ it is clear that for N-type material such that $p_0 \ll N_D$

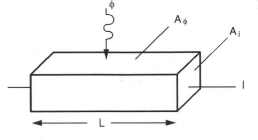

Figure 7.1 The photoconductor detector.

the incremental increase in conductivity (change in σ compared to the dark value) can be written:

$$\Delta\sigma/\sigma_0 = (n'/N_D)(1 + \mu_p/\mu_n) \tag{7.2}$$

We saw in Section 1.3.2 that the energy of photons $E = h\nu$ can create electron–hole pairs by exciting electrons from the valence band up to the conduction band; this occurs only if the photon energy is greater than the band-gap energy E_g of the semiconductor. For silicon, with $E_g = 1.1$ eV, the corresponding wavelength is

$$\lambda_c = hc/E_g = 1.1 \, \mu m \tag{7.3}$$

Above this wavelength the photons do not have enough energy to create electron–hole pairs, and the light passes through the semiconductor device. Below this wavelength the light is absorbed progressively as the photons penetrate below the surface. The absorption law is an exponentially decaying function of depth characterized by the absorption coefficient α, which is a constant for a given wavelength:

$$\Phi(x) = -\Phi_0 \exp(-\alpha x) \tag{7.4}$$

where $\Phi(x)$ is the photon flux density (photons/cm^2); Φ_0 is the surface value. In a photoconductor, for uniform carrier generation the thickness of the bar perpendicular to the incident light must be less than $1/\alpha$.

The speed of response to a change in amplitude of the incident photon flux is governed by the carrier lifetime τ, since we saw in Eq. (1.8) that upon removal of light, the excess carriers decay as $\exp(-t/\tau)$. Since τ is typically of order microseconds in silicon, such a detector has limited applications, although the phenomenon of photoconductivity is extremely useful for the experimental determination of some semiconductor properties. In GaAs, τ is of order nanoseconds, so reasonable but not extremely high speed can be obtained.

To be used as a photodetector a voltage V_L is applied across the bar, length L, and a dc current flows. The photocurrent I_{ph} is given by $I_{ph} = I_{dc}(\Delta\sigma/\sigma_0)$. The photoconductor gain G_{pc} is sometimes used as a figure of merit. G_{pc} is defined as the number of electrons exiting per second across the area A_i in Fig 7.1, e_i divided by the number of electrons generated per second by the photon flux, e_{ph}

$$G_{pc} = e_{ph}/e_i \tag{7.5}$$

In steady state, the generation rate of excess carriers is equal to the recombination rate R [Eq. (2.77)]. Since $R = n'/\tau$, where τ is the carrier lifetime, it follows that for the complete volume of the semiconductor bar $V_{bar} = A_i L$ the number of electrons generated per second is given by:

$$e_{ph} = n' V_{bar}/\tau \qquad (7.6)$$

The time taken for electrons to pass through the bar of length L, is

$$t_{tr} = L/v_n \qquad (7.7)$$

where v_n is the carrier velocity ($= \mu_n E$). In the same way that we defined the channel transit time for the MOSFET (5.44) and JFET (6.42a) this may be expressed as:

$$t_{tr} = L^2/\mu_n V \qquad (7.8)$$

It follows that the total number of electrons per second contributing to the photocurrent I_{ph} is:

$$e_i = n' V_{bar}/t_{tr} \qquad (7.9)$$

By combining Eqs. (7.6) and (7.9), we see that the photoconductor gain may be written as:

$$G_{pc} = \tau/t_{tr} \qquad (7.10)$$

Gains of several hundred may easily be obtained, but only for large carrier lifetimes, that is, only for low-speed detectors.

7.2.2 The P^+NN^+ photodiode

The photodiode is constructed with the contact to the top surface as shown in Fig. 7.2 to allow the light to get through. This ring metallization pattern is frequently used.

Figure 7.3 illustrates the impurity profile of such a P^+NN^+ photodiode (see Chapter 3), with the addition of a curve representing the absorption of photons according to Eq. (7.4) from

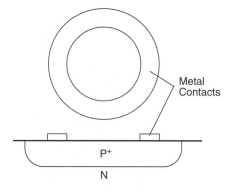

Figure 7.2 Photodiode top surface and cross-section diagrams showing ring metal contact.

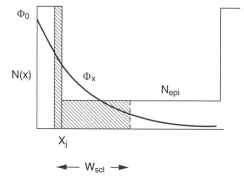

Figure 7.3 P^+NN^+ photodiode: simplified impurity profile $N(x)$ and photon absorbtion law. The depletion layer is shown shaded. (See also CD-ROM BIPGRAPH example PHOTO or PHOTO2; "Minority carrier conc. vs depth.")

incident light perpendicular to the surface. If the diode is reverse biased, the small reverse saturation current I_0 of Eq. (3.25) is augmented by a "photocurrent" I_{ph}. This additional current is made up of three components: (1) electrons generated in the P^+ neutral region diffusing back to the space-charge layer, (2) electrons and holes generated within the space-charge layer traveling in opposite directions due to the electric field, thus creating a drift current, (3) holes generated in the N neutral region diffusing back to the space-charge layer.

These three components add together to generate a total photocurrent I_{ph}. The maximum possible value of I_{ph} (assuming that all incident photons contribute to the photocurrent) is $qA\Phi_0$. In general we can write:

$$I_{ph} = \eta_{opt}qA\Phi_0 \tag{7.11}$$

where η_{opt} is the detection efficiency, or in terms of incident optical power W_{opt}:

$$I_{ph} = \eta_{opt}W_{opt}q/h\nu \tag{7.12}$$

Carriers generated more than a diffusion length from the space-charge-layer boundaries will contribute little to the total current. However, in communications applications (e.g., fiber-optic detectors) speed is of paramount importance, and the thickness of the N region W_n is usually significantly less than a diffusion length. The time taken for the neutral region carriers to diffuse to the space-charge-layer boundaries will be of the order of a diffusion transit time, which from Eq. (3.53) is equal to $W^2/2D$, where W is some thickness less than the total thickness W_n. For $W_n = 10\,\mu$m this gives a time of order 20 ns.

Let us now consider a diode sufficiently reverse biased for the space-charge layer to extend a distance W_{scl} equal to the thickness W_n of the lightly doped region N_{epi}, so that most of the photons are absorbed inside the space-charge layer. The maximum detection efficiency in this case (neglecting loss due to carriers generated in the surface diffused layer) may be obtained by integrating Eq. (7.4). This gives (see Problem 2):

$$\eta = [1 - \exp(-\alpha w_{scl})] \tag{7.13}$$

The time taken for the carriers to drift across the space-charge layer is of order:

$$t_{scl} = W_{scl}/v_s \tag{7.14}$$

where v_s is the saturated carrier velocity [as in the base–collector space-charge layer of a BJT, Eq. (4.66)]. For $W_{scl} = 10\,\mu$m, this gives a time of order 0.1 ns.

The speed of the photodiode responding to a modulated incident light is determined essentially by the above transit times (the circuit time constants determined by the diode capacitance may be calculated separately), and a high-speed photodiode will be designed to have most of the light absorbed within the space-charge layer. In this case the 3 dB bandwidth is given by [1]:

$$B_{3\,db} = 2.4/2\pi t_{scl} \tag{7.15}$$

Combining this fact with the requirement that a large fraction of the incident photons are absorbed leads to the conclusion that a photodiode will normally be designed to have a thickness of order $1/\alpha$ and a background doping level low enough to allow this region to be completely occupied by the space-charge layer at a reasonable reverse bias. Table 7.1 gives α for silicon.

TABLE 7.1
Absorption coefficients for silicon versus wavelength, From [1].

Wavelength, λ (μm)	1.0	0.83	0.73	0.62	0.42
Absorption coefficient, α (cm^{-1})	160	960	2000	4600	50000

The low background doping means that the diode is often referred to as a *PIN* diode photodetector. Figure 7.4 shows plots of quantum efficiency versus wavelength for different thicknesses of the diode lightly doped region. The fall-off in efficiency at low wavelengths is determined by the surface diffused layer (where most of the high-energy incident photons are absorbed).

For example, consider a silicon P^+NN^+ photodiode used as a detector of light emitted by a GaAs light-emitting diode (LED) at 0.85 μm. The photon absorption coefficient at this wavelength is approximately 900 cm^{-1}. A reasonable choice for the N-region thickness would thus be a value greater than 11 μm, say, $W_n = 30 \mu$m. For a doping level of 10^{14} cm^{-3}, Eq. (2.61) leads to the result that in order to have a space-charge-layer thickness of 30 μm, a reverse bias of approximately 60 V is required. Figure 7.3 indicates that the quantum efficiency will be aproximately 80%. The speed of the photodiode will be limited to approximately $t_{scl} = 0.03/10^7 = 3$ ns.

The sensitivity of a photodiode is a key design parameter. This is the minimum optical power required for a given system signal-to-noise ratio. Figure 7.5 shows the noise equivalent circuit for a photodiode with an incident photocurrent I_{ph} and a load resistance R_L representing the input resistance of an amplifier (see Appendix N). If the light is sinusoidally modulated so that the photocurrent has an amplitude I_{ph}, the rms noise current squared will be $I_{ph}^2/2$.

The noise current squared i_{ns}^2 is the shot noise due to the dc current passing through the diode $[2q(I_0 + I_{ph})B$ where B is the noise bandwidth of the amplifier). The signal-to-noise ratio at the output of the amplifier is:

$$S/N = (I_{ph}^2 R_L/2)/[2q(I_{ph} + I_0)BR_L + kT_L B] \qquad (7.16)$$

where T_L is used to denote the equivalent noise temperature of the amplifier. For very low-noise amplifiers, the noise performance will be limited by the dark current of the diode I_0. As seen in Chapter 3, in silicon diodes this is normally due to space-charge generation, Eq. (3.39b).

A technique widely used to minimize the effect of amplifier noise is to use the avalanche multiplication effect to obtain a signal current whose amplitude is multiplied by the factor M [Eq. (3.40)]. The diode is called an avalanche photodiode (APD) and is specially designed to obtain a high-field region for the multiplication and a fast response time (wide bandwidth). Since the avalanche process introduces additional noise, there is an optimum value of M for a given amplifier noise temperature, that is, an optimum reverse bias voltage.

The design of any photodiode will usually involve some trade-off between speed, detection efficiency, and noise performance. Also, it should be clear from the above that the material

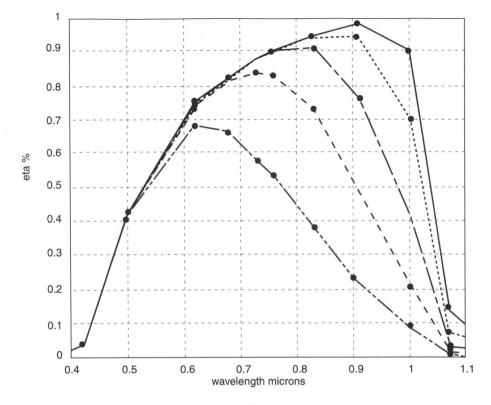

Figure 7.4 Spectral response of internal quantum efficiency η_{opt} of silicon N^+PP^+ photodiodes with center-layer thicknesses of 10, 20, 40, 80, and 160 μm. In each case the doping level and bias are set to ensure that the depletion layer occupies the entire center-layer thickness. The junction depth is 1 μm in all cases. Plots obtained using the BIPOLE3 program.

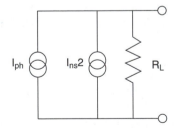

Figure 7.5 Noise equivalent circuit of a photodiode detector with load resistance R_L.

used for the photodiode must be chosen so that its band gap is compatible with the wavelength of the light to be detected. For example, long-distance fiber-optic systems use light in the near-infrared part of the spectrum since there is a minimum attenuation at 1.2 and 1.55 μm and near-zero chromatic dispersion (light at different wavelengths having different velocities) at 1.3 μm. A silicon photodiode cannot be used, but a narrower-band-gap material such as germanium is suitable in this case.

7.2.3 Solar cells

The solar cell is similar to the photodiode except that now the sole goal is to generate as much dc output power as possible. In order to understand this concept, let us examine the overall dc characteristics of the photodiode. We can write the total current as:

$$I_{tot} = I_0[\exp(V_a/mV_t) - 1] - I_{ph} \tag{7.17}$$

where I_{ph} is the current discussed above due to photon absorption. As seen from Fig. 7.6, power is generated only when operating in the lower right quadrant (negative current, positive voltage). The amount of power is optimum for only one set of values of I and V, as determined by the load resistance for a given value of I_{ph} often referred to as the *short-circuit current*. The *open-circuit voltage* V_{oc} is also a frequently quoted number for solar cells. From Eq. (7.17) this is given by:

$$V_{0c} = V_t \ln(I_{ph}/mI_0) \tag{7.18}$$

It is clear that the dc available power will be less than the product $I_{ph}V_{0c}$.

Because of the positive voltage generated, the term *photovoltaic* is often used to describe solar cells. Since the diode is *forward biased* when generating dc electrical power, the space-charge layer will be thin and will contribute little to the overall current. Clearly, the values of I_{ph} and V_{0c} for a given incident optical power will be determined from optical absorption in the neutral region of the diode. Furthermore, since the photons are now created by light over a wide optical spectrum, α will not be a constant, but in silicon varies from about 10^5 to 1 cm^{-1} over the photon energy range 3.5 down to 1.1 eV. The sun's spectrum, corresponding to black-body radiation at 5600°C is very broad, peaking with an irradiance of 2000 W m^{-2} μm^{-1} at a wavelength of approximately 0.5 μm and falling off to one-tenth this value at wavelengths of approximately 0.2 and 1.6 μm (photon energies of 6 and 0.75 eV, respectively).

In order to obtain current from photon absorption at the longer wavelengths, a thick moderately doped N_B background region is required. Following the discussion of photodiodes, the carrier diffusion length must be comparable to or greater than this N_B region thickness, thus necessitating a large carrier lifetime. For example, at approximately 1.1 μm wavelength,

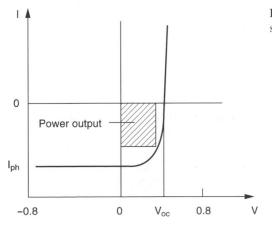

Figure 7.6 *I–V* characteristics of a solar cell showing the power output rectangle.

the value of α in silicon is 100 cm^{-1}. To recover 90% of the incident photons at this wavelength in silicon requires a 230 μm thick layer. Since $L = \sqrt{D\tau}$, the carrier lifetime must be several microseconds. Furthermore, the background doping N_B must not be too low, otherwise series resistance will reduce the available electrical output power. A high-efficiency silicon photodiode will typically have N_B between 10^{16} and 10^{17} cm^{-3} and a thickness of 100 to 200 μm. The diffused layer at the back eliminates a loss of photogenerated current by diffusion to the back contact and gives a slightly higher efficiency than would otherwise be the case.

There is also a limit at the other (high-photon-energy) end of the spectrum. For the impurity profile of Fig. 7.3, where the surface P^+ layer is shown for simplicity to have constant doping, we saw in Chapter 4, with reference to the current gain equation (4.19), that minority carrier diffusion lengths for doping levels greater than 10^{20} cm^{-3} fall below 1 μm. This means that photons absorbed closer to the surface than 1 μm from the junction will not contribute to photocurrent. It might be thought that reducing the junction depth would solve this problem, but surface recombination then dominates. The net result is that the spectral response of a P^+NN^+ solar cell is limited at both the upper and lower ends of the sun's spectrum.

For a well-designed silicon solar cell in bright sunlight on the earth's surface, the value of I_{ph}/A (where A is the diode area) is approximately 0.05 A/cm^2, and the value of V_{0c} is close to 0.6 V in silicon.

Efficiency and choice of material

The efficiency of a solar cell is defined as the dc output power divided by the incident optical (sunlight) power. Since the sun's spectrum extends well into the infrared region, the 1.1 eV band gap of silicon means that all the energy corresponding to wavelengths above 1.1 μm will be lost. This implies that a material with a lower band gap would give better results. However, the requirement for a high open-circuit voltage V_{0c} (a low I_0) necessitates a higher-band-gap material [in Chapter 3 we saw that in all diodes, the reverse saturation current I_0 is proportional to n_i^2 and from Eqs. (2.4) and (2.5) this means a larger value of E_g]. Clearly there is an optimum band gap for a high-efficiency solar cell. GaAs, with a band gap of 1.4 eV, gives slightly greater efficiencies than silicon. Maximum quoted efficiencies for single solar cells are approximately 22% for silicon and 28% for GaAs. Lower-efficiency solar cells may be made using polysilicon instead of crystalline silicon. In this case the efficiency is typically lower than 12%. Amorphous silicon deposited on a glass substrate can be used to make solar cells at even lower cost and still lower power conversion efficiency. The main factor to consider is the overall panel area of the solar cell array for a given power output.

Contacts

The metal contacts to solar cells must be carefully designed to let most of the incident light through but yet keep the contact resistance low. A common configuration is a "comb" structure, as shown in Fig. 7.7. The metal area must be kept well below 50% of the total area so as not to lose too much of the incident light. The series resistance value will be a function of the pattern dimensions and the horizontal current flow in the top diffused layer of the diode characterized by its sheet resistance. It must be kept at a low value to avoid additional power loss.

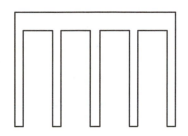

Figure 7.7 Metal contact arrangement for a solar cell.

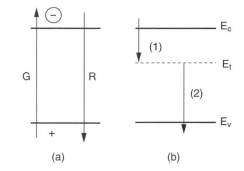

Figure 7.8 Illustration of generation and recombination (a) directly across the band gap, (b) via intermediate levels.

7.2.4 Light-emitting diodes

The concept of photon emission was mentioned in Chapter 1 in connection with the measurement of band gap and recombination and again in Chapter 2 when discussing the mass action law. We pointed out that electrons can be excited from the valence band to the conduction band by high-energy light. We also saw in Chapter 2 that excess carriers (electrons and holes) can be created by injection across a forward-biased P–N junction. In both cases under steady-state conditions, the excess electrons in the conduction band will recombine with the excess holes in the valence band, the process being characterized by the carrier recombination lifetime τ. In the simplest case shown in Fig. 7.8(a) the recombination occurs directly across the band gap E_g with release of energy. In the case of "direct radiative recombination" this energy will be in the form of photons with an energy $h\nu = E_g$. This is the case for materials such as GaAs. In silicon the process is more complicated and occurs mainly through recombination centers near the middle of the band gap, as shown in Fig. 7.8(b). In this case the energy is largely converted to phonon (heat) energy with little or no photon emission.

The generation of light is therefore limited by the choice of material. Since GaAs has a band gap of 1.4 eV, it will radiate in the near-infrared part of the spectrum at 0.89 μm. This cannot be seen by the naked eye. If some phosphorus is added to make GaAsP, the band gap can be increased to produce red light. Other combinations of materials are possible to produce light-emitting diodes (LEDs) with various light wavelengths in the visible part of the spectrum.

A P–N junction made with such material and forward biased to create minority carrier injection will thus emit light over a narrow band of frequencies (monochromatic light) determined by the band gap. The intensity of the radiated light is proportional to the concentration of excess carriers and therefore to the current. The efficiency (light power out divided by electrical power consumed) is a function of the generation–recombination process and internal absorption and reflection loss. Of the three components of current discussed in Chapter 3, electron injection current I_n [Eq. (3.8)], hole injection I_p [Eq. (3.7)], and space-charge recombination I_{scr}

[Eq. (3.33)], only the first normally contributes to light emission in an LED. We can thus define a current injection efficiency [2]:

$$\gamma = I_n/(I_n + I_p + I_{\text{scr}}) \tag{7.19}$$

An internal quantum efficiency η_q can also be defined as the fraction of electrons that recombine with light emission compared to the total number that recombine. In the simplest case the radiative recombination characterized by a carrier lifetime τ_R occurs from the conduction to the valence bands, an energy difference of E_g as shown in Fig. 7.8(a). Nonradiative recombination with a lifetime τ_{NR} also occurs from recombination centers at intermediate energy levels, as in Fig. 7.8(b). The respective recombination rates are:

$$U_R = n'/\tau_R \tag{7.20}$$

$$U_{NR} = n'/\tau_{NR} \tag{7.21}$$

The corresponding quantum efficiency is thus:

$$\eta_q = U_R/(U_R + U_{NR}) = 1/(1 + \tau_R/\tau_{NR}) \tag{7.22}$$

This is a function of doping level and of the parameters describing the physical recombination mechanisms. A maximum value is typically of order 4% for a GaP diode. The overall internal efficiency η_i is given by [2]:

$$\eta_i = \gamma \eta_q \tag{7.23}$$

This is greater than the external efficiency because of the absorption and reflection of emitted light inside the device. A significant part of the design of LEDs involves interfacing the semiconductor diode to the "outside world". The interface is chosen to minimize loss due to internal reflection. By choosing a material with a refractive index between that of the semiconductor and air and by shaping the encapsulation appropriately, overall efficiencies of order 1% are achievable. Figure 7.9 shows a typical encapsulated LED using epoxy in a hemispherical shape.

It may be noted that the 0.85 μm emission from a GaAs diode is well matched to the detection peak of silicon photodiodes. This provides a convenient method of transmitting data over short fiber-optic digital links, including the optical isolator to couple electrical signals without any electrical interconnection.

7.2.5 The semiconductor *PN* junction laser

The light emitted by a forward-biased GaAs diode will be noncoherent; that is, the spectrum will spread over a few hundred angstroms. It is possible to obtain coherent emission of light over

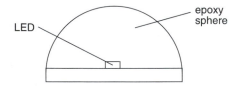

Figure 7.9 Light-emitting diode showing typical epoxy encapsulation.

a very narrow spectrum (of order 1 or 2 Å) if the property of light amplification by stimulated emission of radiation (LASER) is invoked and made possible by using specially constructed *PN* diodes. The essential property concerns electron–photon interaction. In the previous discussion of LEDs, it was assumed that when an electron (excited into the conduction band by a forward bias across the *PN* junction) dropped down to the valence band, a photon of energy $h\nu = E_g$ was emitted. If, however, this photon is constrained within the diode, retaining its energy, it is possible for it to interact with an already excited electron with generation of a second photon having the same energy. If the phase of the light corresponding to each of the two photons is the same, a build-up of light will occur. This is the stimulated emission concept, and if the process continues, amplification (increase in intensity) of the original light will occur.

The necessary conditions can be met by using a P^+N^+ diode (high doping levels on both sides) and applying sufficient forward bias so that electron build-up occurs due to injection in the *P* side close to the junction, with empty energy states at the top of the valence band. This situation is referred to as *population inversion* and is shown in Fig. 7.10.

For doping levels above approximately 10^{19} cm^{-3} on both sides of the junction, the Fermi level is above E_C on the *N* side and below E_V on the *P* side [see Appendix F, Eq. (F.20)]. If a forward bias V_a greater than E_g/q is applied, the situation shown in Fig. 7.10(b) is obtained, where electrons injected from the *N* side reside above E_C on the *P* side. At the same time, because the Fermi level on the *P* side is below E_V, there are a large number of available energy levels between E_{FP} and E_V. Thus the required conditions are satisfied for the creation of photons by the already excited conduction-band electrons in a narrow region close to the junction. It is in this region that laser action occurs.

The requirement that the created photons be available to create new photons from these electrons is made possible by constraining a large fraction of the light within the diode. This is done by making highly reflecting walls on two opposite sides, forming a Fabry–Perot cavity. To avoid multiple-mode lasing action, the other four sides are roughened. The reflecting sides can be created by cleaving the crystal; the refractive index of the material (e.g., $n = 3.6$ for GaAs) ensures good reflectivity. A simple picture of a GaAs laser is shown in Fig. 7.11, where the vertical *x* direction is marked and is the same as the *x* direction marked on Fig. 7.10.

As the forward bias is increased, initially noncoherent light will be generated. At a threshold current corresponding to the start of the creation of the population-inversion condition (a forward bias slightly greater than E_g/q), laser action starts, and the light becomes coherent monochromatic radiation. The light is radiated as shown in a narrow region on the *P* side,

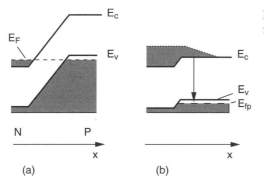

Figure 7.10 P^+N^+ laser diode (a) zero bias; (b) forward bias showing population inversion.

parallel to the junction, with enough photon energy being constrained within the cavity walls to create the laser action. A wide range of direct radiative recombination materials is available for *PN* injection lasers, including GaAs, InSb, and InP. Power conversion efficiencies can be high, of order 50%. Current densities are high (corresponding to the very high forward bias) and for continuous operation (i.e., nonpulsed) careful heat removal (use of heat sinks) and/or cooling are necessary.

One disadvantage of the above structure is that the laser action only occurs in a small region close to the junction, but the electrons diffuse and recombine up to several diffusion lengths away from the junction. This is inefficient because of the "wasted" current. Performance improvement can be obtained by constraining the region for light emission. This can be accomplished by the addition of a region of higher band gap a short distance from the junction, as shown in Fig. 7.12.

For a GaAs laser this can be done by using GaAlAs, whose bandgap increases with the Al fraction, x, according to the law:

$$\Delta E_g = 1.247x \tag{7.24}$$

For the electrons, this has the same effect as a low–high junction (Appendix J) and both prevents the diffusion of electrons and keeps the current at a lower level for a given applied bias. Furthermore, the refractive index of the larger-band-gap material is lower, and this prevents the escape of photons outside the active region. This is the *heterojunction laser*, and the design leads to higher efficiencies with less heat generation (operation at lower current densities). Further improvements can be obtained using a *double-heterojunction* system if a wider-band-gap N region is also used.

7.2.6 Imaging arrays using charge-coupled devices

The charge-coupled device (CCD) is used for various nonoptical applications (e.g., delay lines). We include it here because of its increasingly widespread use for optical imaging arrays. The operation of the CCD depends on what happens when a step voltage V_G is applied to a MOS capacitance, with a signal charge Q_S present.

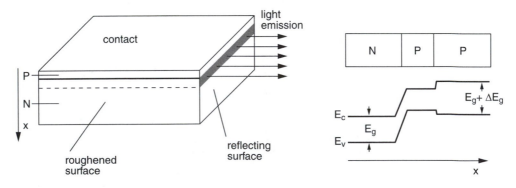

Figure 7.11 Simplified drawing of a *PN* junction laser.   **Figure 7.12** The heterojunction laser.

Consider the MOS capacitance of Section 5.2 with a P substrate and a positive voltage step V_G applied at time $t = 0$. Because the time taken (referred to as the thermal relaxation time) for an inversion layer to form is of order seconds, a depletion layer is formed almost instantaneously. This is called *deep depletion*. The surface potential V_S is given by:

$$V_G = V_{FB} - Q_S/C'_{0x} + V_S \tag{7.25}$$

where in the case of the CCD Q_S is made up of the depletion-layer charge (4.2) and a signal charge Q_{sig}.

$$Q_S = -qN_Ad_p - Q_{sig} \tag{7.26}$$

where $V_S = qN_Ad_p^2/2\epsilon$ and where all this voltage is developed across the depletion layer for the short times under consideration. Combining the above gives:

$$V_G - V_{FB} = (2\epsilon qN_A V_S)^{1/2}/C'_{0x} + Q_{sig}/C'_{0x} + V_S \tag{7.27}$$

Defining:

$$V_A = V_G - V_{FB} - Q_{sig}/C'_{0x}$$
$$V_B = \epsilon qN_A/C'^2_{0x}$$

and solving for V_S gives:

$$V_S = V_A + V_B - (2V_A V_B + V_B^2)^{1/2} \tag{7.28}$$

Thus for a given gate voltage and signal charge, the surface potential, and hence the depletion-layer characteristics, are known.

The operation of a CCD may be understood by considering the basic properties of the metal oxide semiconductor system discussed in Section 5.2. Figure 7.13 shows the basic CCD system. A series of gate metal contacts are biased at potentials V_2, V_1 as shown. Let us consider the situation shown at time $t = t_1$ with a positive potential step V_2 greater than V_1 applied to gate 1. Upon application of this step V_2 (assuming P-type silicon substrate), the depletion layer forms with Eq. (7.28) giving V_S. A charge of electrons is then optically generated. The electron charge will remain at the surface due to the large potential V_2.

If now the voltages on the gates are altered to those shown at time $t = t_2$, with a larger potential on the following gate electrode 2, the electrons will flow toward the lower-energy condition on the right (by a combination of diffusion and drift). At time $t = t_3$ gate 3 is raised to the high potential with gates 1 and 2 at the lower potential so the charge is transferred to gate 3. The charge can thus be transferred (or coupled) progressively along a long line of CCD elements. The charge on a final element can be sensed by monitoring either the value of V_S or the capacitance, since $Q_B = (2qN_A\epsilon V_S)^{1/2} = qN_Ad_p$, where d_p is the depletion-layer thickness, which determines the value of the capacitance as in Eq. (5.8).

In a practical circuit the voltages are formed by clocked pulses at frequencies up to over 100 MHz. Some charge will be lost between each transfer. If each cell is a distance L from the next cell, the time taken for transport by diffusion will be related to the diffusion transit time Eq. (3.53): $t_{tr} \propto L^2/D_n$. The repulsive force between electrons will assist charge transfer as will carrier drift due to fringing fields. In typical CCD arrays a transfer efficiency of 99.99% is obtained at clock frequencies of several MHz.

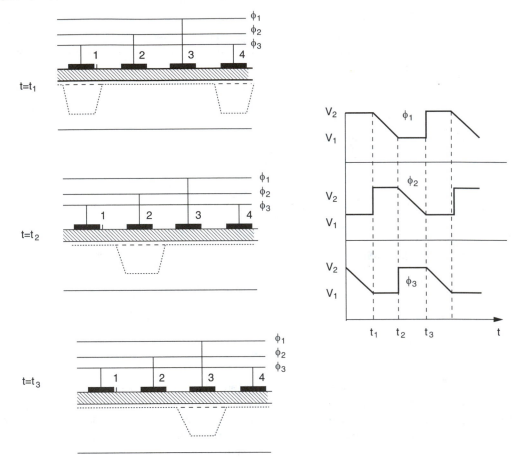

Figure 7.13 A three-phase CCD array showing electron transfer by application of suitable voltages to the gate electrodes. The substrate is *P* type, and the dotted region shows the depletion layer (after Tyagi [5]).

The CCD array is used for two main purposes: (1) as a delay line or shift register, (2) as an imaging array (for example, in still and video camera applications). In the imaging array application, a two-dimensional CCD array is created with the optical image being incident on the surface. The light intensity and integration time (set by the clock frequency) on a particular cell determines the electron charge on that cell. By transferring the charge on each cell progressively to the end element and sensing the value of each charge as it reaches the end element, a signal level proportional to the image intensity on each cell or pixel is obtained. This can then be done on each successive line of a two-dimensional CCD imaging array to build up a picture (in a manner similar to TV scanning).

Modified versions of the above include reduced spacing by using polysilicon gates. Buried channel CCDs can also be made using an ion-implanted *N* layer between the gate oxide and the *P* substrate. This gives improved charge-transfer efficiencies. Two-phase CCD arrays (which simplify interconnect layouts because there is one fewer line) can be built by incorporating directional transfer using stepped oxides or a combination of polysilicon and metal gates.

7.3 HIGH-POWER BIPOLAR DEVICES

7.3.1 High-power bipolar transistors

The high-current high-voltage BJT structure differs in two aspects from the conventional small-geometry structure discussed in Chapter 4. First, in order to obtain very high-voltage breakdown, the collector must be lightly doped and very thick; for example, from Eqs. (4.40) and (4.46), for $BV_{ceo} = 500$ V, the collector will be doped at a level of order 10^{14} cm^{-3} and will be over 50 μm thick. This means that frequently the background doping of the wafer will form the collector region (as opposed to having a moderately doped epitaxial layer grown on the top surface). Second, there is a severe constraint on the emitter dimensions. One cannot simply make a large square emitter in order to obtain the large area necessary for high-current operation. This restriction follows from the discussion of emitter current crowding in Section 4.5, Eq. (4.33), where the base current I_b must be limited to not much higher than $0.025/r_{bb}$. From Eq. (4.22), if the emitter width L is increased at the same time as its other dimension B, r_{bb} remains constant, and hence the maximum value of I_B (and hence the maximum value of $I_C = \beta I_B$) remains constant. The way around this restriction is to make an interdigitated layout consisting of a number n elemental emitters in parallel, each one having a large B/L ratio, as shown in Fig. 7.14.

If B_s is the individual stripe length, then the total active area (neglecting end effects) is:

$$A = nLB_s \qquad (7.29)$$

rewriting Eq. (4.22) gives the base resistance as:

$$r_{bb} = [\rho_b L/(12W_b B_s)]/n \qquad (7.30)$$

The area has been increased by n and the base resistance decreased by $1/n$, thus allowing very large areas to be obtained with low base resistance. The maximum length B_s is limited by metal contact resistance in the B_s direction because of the voltage drop $I_E R_E$ due to the emitter current and the resistance $R_{MSQ}(B_s/L)$, where R_{MSQ} is the metal sheet resistance (determined by its resistivity and thickness).

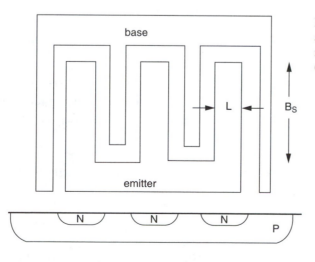

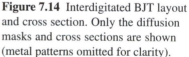

Figure 7.14 Interdigitated BJT layout and cross section. Only the diffusion masks and cross sections are shown (metal patterns omitted for clarity).

The above layout can be implemented in either discrete form or in an integrated circuit. In the former case the transistor can be several millimeters or more in size.

7.3.2 The Darlington pair

The Darlington pair originated as a circuit combination shown in Fig. 7.15(a). This arrangement provides a current gain (which is the product of the individual gains $\beta_1\beta_2$) and a correspondingly high input impedance. It is particularly easy to fabricate on a single substrate because the two collectors are connected in the actual circuit. A cross-sectional view of the complete structure is shown in Fig. 7.15(b). The "driver" transistor T_1 will operate at a current that is $1/\beta$ of the output transistor T_2. The areas of the two transistors should therefore differ by approximately this ratio to maintain similar current densities. The cross-section diagram in Fig. 7.15(b) indicates an interdigitated structure for T_2 and a single emitter for T_1, although both transistors may be interdigitated for very high-current applications. Currents up to 100 A at 600 V have been handled using large-area Darlington pairs. It may be noted that in high-voltage power applications the transistors are usually operated in heavy saturation; the individual current gains are therefore well below the normal active regime values.

The resistors shown in Fig. 7.15(a) may be integrated as part of the base regions of T_1 and T_2 and are necessary in power switching applications during "turn-off" in order to evacuate the charge.

7.3.3 The silicon-controlled rectifier or thyristor

For very high-voltage applications the silicon-controlled rectifier (SCR) or thyristor finds widespread use. This is a four-layer device made up of two overlapping bipolar transistor structures. It is represented in Fig. 7.16 by its circuit model and by its doping profile.

The collector of the *NPN* transistor is the base of the *PNP*, and the base of the *NPN* forms the collector of the *PNP* device. A circuit analysis of this arrangement leads to the fact that the current between cathode and anode becomes infinite if the condition:

$$\alpha_{NPN} + \alpha_{PNP} = 1 \tag{7.31}$$

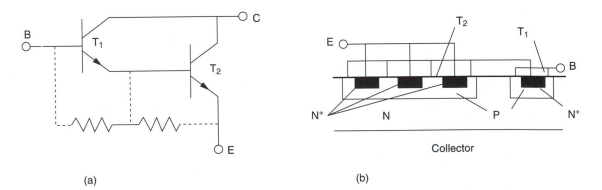

(a)

(b)

Figure 7.15 (a) BJT Darlington pair; (b) cross section of BJT Darlington pair.

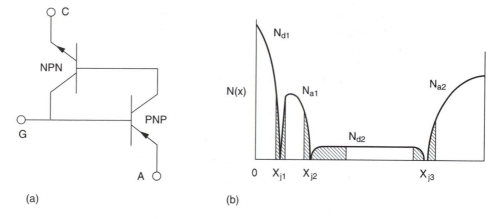

Figure 7.16 (a) Circuit representation of the SCR; (b) doping profile of *NPNP* SCR structure.

where α is the common-base current gain [$\alpha = \beta/(1+\beta)$]. We saw in Chapter 4 that the current gain of a BJT falls off at low V_{BE} bias (this falloff can be enhanced by placing a resistance between emitter and base terminals). This property enables the SCR to be used as a switch by using the gate terminal to control the operating point. The actual current–voltage characteristic is as shown in Fig. 7.17 for the case where the gate is open circuit. Let us examine the operation as the applied voltage is increased. On the bottom part of the characteristic the current is the reverse leakage current of the junctions, with a slight bias dependence due to space-charge generation (see Section 3.6). As the voltage increases, so the depletion layer of the second (X_{j2}) junction increases in thickness. This reduces the neutral base width of the *PNP* transistor, and hence its current gain increases as predicted by Eq. (4.39). The gain of the *NPN* transistor increases because of a slight increase in forward bias across X_{j1}. Eventually the *PNP* base width becomes very narrow and its gain increases rapidly; this is for bias voltages close to the "turn-on" voltage V_{on}. The current starts to rise, and the gain of the *NPN* transistor increases (due to a reduced contribution of space-charge recombination current as in Fig. 4.19). Eventually the

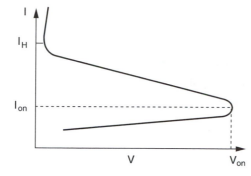

Figure 7.17 Two-terminal *I–V* characteristic of an SCR (gate open).

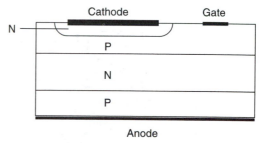

Figure 7.18 Cross-sectional view of an SCR.

gain of the *NPN* transistor is sufficiently high that the condition $\alpha_{NPN} + \alpha_{PNP} = 1$ is satisfied for increasingly large *PNP* base widths, that is, for decreasing values of applied bias. This is the negative slope region of the characteristic. As the "holding current' I_H is approached, the center depletion layer around X_{j2} shrinks, and this junction eventually becomes forward biased. Once the current exceeds the value I_H, the SCR remains in the "on" condition. This two-terminal operation is referred to as a *Shockley diode*. Addition of a gate contact (base of the *PNP* transistor) enables the "on" condition to be governed by the signal applied to the gate terminal.

A cross-sectional view of the SCR is shown in Fig. 7.18. The high current passes vertically from cathode to anode, and this area occupies most of the surface of the device. Various surface layouts are used, including circular or spiral geometries. A major part of the design concerns transient behavior. Because of the thick *N* region, a substantial amount of excess carrier charge builds up. This is the determining factor in the switching speed. For a 1000 V SCR, speeds of order tens of microseconds are typical.

7.4 HIGH-POWER MOS DEVICES: DMOS AND VMOS

The interdigitated layout arrangement used for high-current bipolar transistors cannot be used to increase the current-handling ability of conventional MOS devices because in the latter case three terminals are necessary on the surface, as opposed to only two for the BJT (the collector being contacted at the bottom). To circumvent this problem some alternative MOS structures have been developed. We shall examine briefly two such devices: DMOS and VMOS. In both cases the goal is to have only two contacts (gate and source) on the top surface, with the drain being the bottom contact.

7.4.1 The vertical DMOS structure

Figure 7.19 shows a cross section of a vertical double-diffused MOS or DMOS structure.

The *P* diffusion is analogous to a bipolar transistor base diffusion, with the N^+ drain formed like an emitter diffusion. The surface of the *P* diffusion under the gate is the channel region,

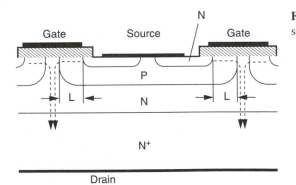

Figure 7.19 Cross section of vertical DMOS structure (e.g., see [10, 11]).

length L. The source–drain current flows horizontally as in a normal MOSFET, until the N region is reached, when the current turns to the vertical direction due to the application of a drain voltage to the bottom contact. The channel length L is determined by the sideways diffusion of the P and N dopants. Now that only two contacts are on the top surface, various geometrical configurations may be used for interconnection of a large number of active channel elements. In practice the gates are made using doped polysilicon instead of metal (see Chapter 8 for details). This enables an oxide layer to be grown above the polysilicon contact, thus providing an extra degree of freedom in the layout. In the HEXFET structure, for example, the polysilicon gates are hexagonal, with a large number interconnected. In this manner very high currents may be controlled by a moderate gate voltage.

7.4.2 The vertical VMOS structure

Figure 7.20 shows a cross section of another vertical MOS device. The V groove is obtained using the fact that KOH etches preferentially along certain crystal axes; it is an anisotropic (i.e., nonuniform) etch. On a $\langle 100 \rangle$-oriented substrate, KOH etches at 54.7° to the horizontal plane. The P and N diffusions on an N epitaxial layer create an N (source), P (channel region), N drain with the channel length L being now as shown at an angle. If L is visualized as being vertical, operation immediately becomes clear; the MOSFET behaves just like a normal N-channel P-substrate device with the source–drain current flowing vertically. A single large drain contact on the bottom is used. This structure thus provides a layout with only two contacts on the top surface, thus enabling several possibilities for interconnect layouts with no fundamental constraint on the number of paralleled devices.

The above two structures show that paralleling of MOS devices is possible, although it must be recognized that the source–drain current is always constrained to be in the silicon just inside the oxide–silicon interface. Although the drain current becomes vertical, the current density is high only over a very small area (under the active device), and the total drain current per unit drain area cannot be as high as in the interdigitated BJT. Nevertheless, the high-power MOSFET structures have the distinct advantage of infinite input (gate) resistance and for this reason find widespread use.

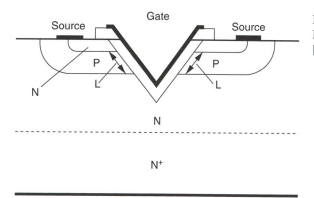

Figure 7.20 Cross section of vertical MOS (VMOS) structure. After Bar-lev [10] and Holmes and Salama [12].

7.5 THE INSULATED GATE BJT

The essential features of the insulated gate bipolar transistor are shown in Fig. 7.21. This structure combines the advantages of the high-voltage high-current BJT with the high input impedance of a power MOS device. On the left is a conventional vertical *PNP* transistor. A vertical parasitic *NPNP* thyristor is next. Under the gate (G) is the *P* background for the *N* channel of a structure very similar to the DMOS discussed above. The operation of this composite device can best be understood with the aid of an equivalent circuit shown in Fig. 7.22. The description given below follows closely the work of J. Baliga [4], to which the reader is referred for additional information about this extremely useful structure. The device is essentially a vertical *PNP* BJT with the base controlled from the source–drain current of a MOSFET.

When a negative voltage is connected to the collector C, the P^+N junction J_1 is reverse biased, with the depletion layer extending upward into the N^- region. Only reverse leakage current flows, and eventually breakdown occurs (see Fig. 7.23).

When a postitive voltage is connected to the collector C, with zero volts between gate and emitter, the top junction J_2 becomes reverse biased, and no current flows. This is the forward blocking region. As V_{CE} is increased, a depletion layer forms around junction J_2 in the N^- region, and only reverse leakage current flows, followed eventually by breakdown.

Consider now a positive voltage applied to the gate such that an inversion layer forms in the *P* region of the MOSFET channel. The source–drain current provides the base current for the *PNP* BJT, thus allowing it to turn on, with a forward-biased emitter–base junction J_1. The P^+ emitter injects holes into the wide N^- base, eventually forcing it into high-level injection (see Section 4.5). The voltage drop in the N^- region can thus be kept very low (as in a *PIN* diode, see Section 3.3) and the IGBT is in its forward-conducting region. If the gate voltage is reduced to a value close to the threshold voltage of the MOSFET (such that the horizontal channel voltage drop is equal to the difference between the gate and threshold voltage), the pinch-off condition occurs and the drain current is limited. The base current of the BJT is thus limited and hence so is the collector current. This is the active region where the current remains almost independent of V_{CE} as in the active region of a BJT. Turnoff can be accomplished simply by

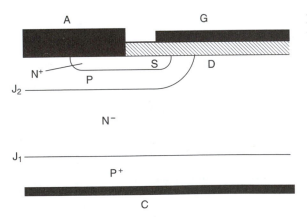

Figure 7.21 Insulated gate bipolar transistor (IGBT) cross-sectional view. After Baliga [4].

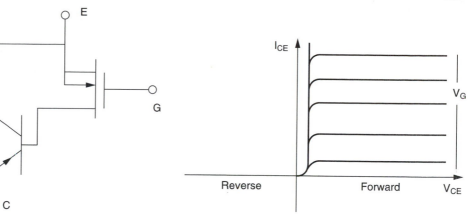

Figure 7.22 Equivalent circuit of IGBT.   **Figure 7.23** dc characteristics of IGBT.

reducing the gate–emitter voltage to zero. There is the usual delay associated with the excess charge storage in the N^- base region as in the collector of a BJT (see Section 4.8).

The IGBT combines the advantages of a large vertical current for the vertical BJT, with the purely capacitive input impedance of a MOSFET. It finds wide practical applications for high-voltage high-current switches.

7.6 MICROWAVE DEVICES

7.6.1 Microwave *PIN* diodes

The microwave *PIN* diode is identical in structure to that presented in Section 3.3. However, in a high-voltage rectifier application, the center I layer is of order 100 μm thick and with diameters that can be several centimeters for high-current diodes. The microwave device, on the other hand, is usually designed for low voltages (< 10 V) with a small cross-sectional area to maintain low capacitance; its main application is as a microwave switch or attenuator. To understand this concept, we refer to Figure 3.6, redrawn in Fig. 7.24 with an ac carrier modulation superimposed on the static carrier distribution.

The dc carrier distribution has the time-varying component superimposed for one instant of time. The amplitude decays toward the center because of the time taken for the carriers to diffuse. The frequency-dependent characteristic decay length is referred to as the high-frequency diffusion length. If the center-layer width is W_i, it takes a time t_{diff} of order W_i^2/D for the carriers to diffuse this distance. For $W_i = 2\,\mu$m, $D = 20$ cm^2/V s, $t_{diff} = 1$ ns. For a microwave signal at 10 GHz, the superimposed modulation will decay quite rapidly. The impedance "seen" by the microwave signal for most of the distance W_i is then given approximately by the resistance due to the carrier concentration, or $W_i/(q\mu n A)$ where μ is a combined electron and hole mobility, n is the bias-dependent carrier concentration (a slight function of depth), and A is the diode cross-sectional area. This resistance is inversely

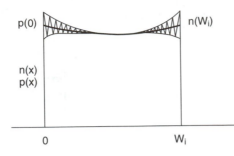

Figure 7.24 Microwave *PIN* diode: carrier concentration distribution in the presence of a sinusoidal microwave signal.

proportional to n and thus to the bias current (3.17). The *PIN* diode may thus be used as a variable attenuator when combined appropriately in a microstrip line.

Alternatively, one may envisage two bias states: (1) heavy forward bias where the above resistance is very small (compared to a 50 ohm microstrip line impedance, for example); (2) reverse bias where the only impedance presented by the diode is its depletion-layer capacitance. The *PIN* diode can thus be used as a microwave switch in series or parallel with an appropriate microstrip circuit. The microwave signal may be blocked or passed according to the applied dc bias.

7.6.2 Varactor diodes and the hyperabrupt junction

The varactor diode is a *PN* junction in which the nonlinear capacitance variation with applied bias voltage is used. The $C(V)$ law of a junction is generally expressed in the form:

$$C(V) = C_0/(1 - V_a/V_{bi})^\gamma \qquad (7.32)$$

Applications include parametric amplification, frequency multiplication, and automatic frequency control (where the capacitance is used as part of an oscillator circuit, with appropriate detection of the signal and feedback to the diode bias network to adjust its capacitance in such a manner as to keep the resonant frequency constant). For an abrupt junction, the depletion-layer capacitance has been shown to be proportional to $1/\sqrt{V_{jtot}}$, where $V_{jtot} = V_{bi} - V_a$ (see Section 3.8.1). In Sections 2.4.6 and 3.8.1 we also pointed out that for a linear graded junction where $N(x) = ax$, a capacitance law $C_j \propto 1/V_{jtot}^{1/3}$ is obtained. γ is therefore equal to 1/2 for an abrupt junction and 1/3 for a linear graded junction. (See also CD-ROM BIPGRAPH example DIOD30; "Capacitance exponent MJ vs V_{jtot}" were MJ(γ) varies from 1/3 to 1/2 as the bias varies from forward to large reveerse values.)

In some applications a nonlinear dependence with γ greater than 1/2 is required. This is possible using a "hyperabrupt" doping profile as shown in Fig. 7.25. For increasing reverse bias, the depletion layer extends from the junction X_j to the region where the impurity profile decreases with increasing x. It therefore becomes progressively "easier" for the depletion layer to penetrate to larger values of x (lower values of capacitance) and so the capacitive nonlinearity increases. In Section 4.6 we saw that C_{jc} for the base–collector junction of a BJT normally has an exponent of 1/2 and C_{je} for zero and forward bias an exponent of 1/3. In fact, if the emitter–base junction is reverse biased, a *hyperabrupt* capacitance–voltage law can sometimes be observed, if the depletion layer extends past the peak of the base doping before junction breakdown occurs. In the varactor diode this type of behavior is required when a strong dependence of capacitance on bias voltage is necessary, as, for example, in tuning or automatic frequency control (AFC) applications.

For frequency multiplication and low-noise parametric amplification, constant doping on the center layer is normal with $\gamma = 1/2$; in this case the optimum design is one for which the cut-off frequency

$$f_c = 1/2\pi R_S C \tag{7.33}$$

is maximum. Here R_S is the diode series resistance. Since both the series resistance and the capacitance of the diode depend on the center layer doping, it can be readily shown that f_c is closely related to the dielectric relaxation time $\tau_{rel} = \rho\epsilon$, where $\rho = 1/q_n N_{epi}$. To a fair approximation (depending on the bias conditions) one can write $f_c = 1/2\pi \tau_{rel}$.

7.6.3 The IMPATT device

IMPATT stands for impact avalanche transit time device. This is a slightly modified *PIN* diode (in fact the *PIN* diode can also exhibit IMPATT properties). The simplified impurity profile of a Read IMPATT diode is shown in Fig. 7.26.

As can be seen, this is really a $P^+NN^-N^+$ diode structure. Under reverse-bias conditions the electric field distribution is established as shown, with a maximum at the junction, where the doping is relatively high on both sides, falling to a low value (with slight negative slope for finite N^- doping level) throughout the N^- region of total thickness d. In operation a microwave signal is superimposed on the dc voltage and the diode is placed across a parallel resonant circuit.

Consider the case where the peak electric field is high enough for avalanche multiplication (impact ionization) to occur (i.e., a field of order 3×10^5 V/cm) at the junction, but lower by a factor 1/2 or so throughout the N^- region. The electrons created by impact ionization will drift to the right at their saturated drift velocity v_s. After a time t_d they will reach the depletion-layer edge. The critical drift transit time is given by:

$$t_d = d/v_s \tag{7.34}$$

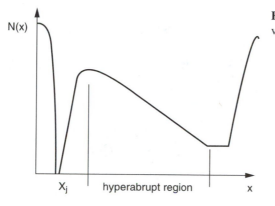

Figure 7.25 Impurity profile of hyperabrupt varactor diode.

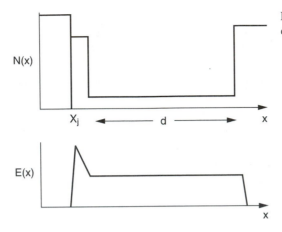

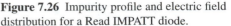

Figure 7.26 Impurity profile and electric field distribution for a Read IMPATT diode.

If this time is made equal to one-half a period of the microwave signal, and since the impact ionization is created at the peak of the ac voltage, then the conduction component of current is out of phase with the voltage. This is a negative resistance situation. This means that whatever original ac current amplitude existed without the IMPATT diode, it will have an increased amplitude due to the IMPATT effect. In other words, amplification of the original signal occurs. In the limit, oscillations can be obtained, and the IMPATT diode with its associated circuit or cavity becomes a source of microwave power.

Example of IMPATT diode calculation:

Consider an IMPATT diode with an N^+ region 0.2 μm thick doped 10^{17} cm^{-3} and an N region 2 μm thick doped 10^{14} cm^{-3}. For a peak electric field $E_{peak} = 4 \times 10^5$ V/cm, calculate the required bias voltage and the oscillation frequency.

In the N^+ region, $dE/dx = (q/\epsilon)N_D^+ = (1.6 \times 10^{-19}/10^{-12}) \times 10^{17} = 1.6 \times 10^{10}$. The field at the edge of the N^+ region is thus $4 \times 10^5 - 0.1 \times 10^{-4} \times 1.6 \times 10^{10} = 2.4 \times 10^5$ V/cm. For the N region, $dE/dx = 1.6 \times 10^7$ V/cm^2; the drop in electric field over the 2-μm thick-layer is thus $1.6 \times 10^7 \times 0.2 \times 10^{-4} = 0.32 \times 10^3$ V/cm which makes the field distribution practically horizontal. The voltage drop across this layer is thus $2.4 \times 10^5 \times 2 \times 10^{-4} = 48$ V. The voltage across the N^+ layer is $[2.4 \times 10^5 + (4 \times 10^5 - 2.4 \times 10^5)/2] \times 0.2 \times 10^{-4} = 6.4$ V. The total bias voltage required is thus 54.4 V.

For a drift region thickness d equal to 2 μm, $t_d = 2 \times 10^{-4}/10^7 = 20$ ps. The corresponding oscillation frequency is thus 25 GHz.

7.6.4 The transferred-electron (Gunn) device

In Chapter 2 we discussed the velocity versus electric field characteristics of semiconductors. In silicon the velocity saturates as in Fig. 2.10. However, in some semiconductors such as

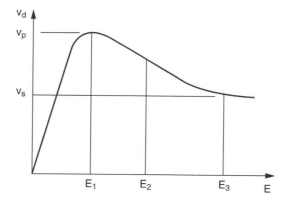

Figure 7.27 Velocity versus electric field diagram for GaAs.

gallium arsenide (GaAs) and indium phosphide (InP) the measured velocity–field diagram exhibits a peak at intermediate values of field, as shown in Fig. 7.27.

For GaAs the values are: $v_p = 2.2 \times 10^7$ cm/s, $v_s = 10^7$ cm/s, $E_1 = 3.2 \times 10^3$ V/cm, $E_3 = 2 \times 10^4$ V/cm. The shape of this curve can only be explained theoretically by a detailed examination of the band structure; this shows that the behavior is due to the transfer of electrons between different energy states in the conduction band, and this category of devices is therefore also referred to as TED (transferred-electron device). These semiconductors exhibit the peak drift velocity v_d followed by a decreasing velocity as the field is increased. Since current is proportional to electric field (for constant carrier concentration), this represents a negative resistance region of the I–V characteristic for bulk material. The Gunn device consists of an N region contacted at both ends through N^+ layers. In operation as a source of microwave power, the device is coupled to a resonant cavity. The combination of negative resistance and cavity allows oscillations to build up due to propagation of a dipole of charge through the device.

Figure 7.28 illustrates this behavior. Assume that the voltage V creates an average electric field such that operation is in the negative slope region of the v_d versus E diagram. This means that the differential mobility is negative. The dielectric relaxation time ϵ/σ determines the rate of decay towards charge neutrality if the charge neutral condition is temporarily upset. A negative mobility denotes a negative conductivity, and instead of a charge decay we get a charge build-up. A charge build-up in the above situation will be accompanied by an increasing

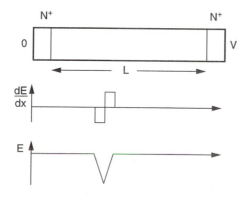

Figure 7.28 Gunn device: (a) structure; (b) idealized space-charge at time t; (c) electric field at time t.

magnitude of electric field, and thus a decrease in electron velocity (i.e., moving toward the point v_s, E_3 in Fig. 7.27). We thus have the situation of higher-velocity electrons moving in from the left to create a buildup of electrons (negative charge) and a higher velocity of electrons leaving from the right, creating a depleted region (positive charge). This dipole of charge will travel to the right (due to the electric field). When it reaches the N^+ contact, a pulse of current will be transferred to the external resonant circuit. The process then repeats with buildup of a new charge diople at the left. The time taken for the charge dipole to cross the device is approximately L/v_s, and this determines the frequency of oscillation (to which the external circuit must be tuned). For $L = 1\ \mu$m, $v_s = 10^7$ cm/s, this gives a frequency of approximately 16 GHz.

Other modes of operation are possible. In the Limited Space-charge Accumulation (LSA) mode the oscillation frequency is controlled by the external circuit. This is possible if the transit time L/v_s is less than the time required for creation of the charge dipole.

7.6.5 Heterojunction bipolar transistors

With the advent of new technologies such as molecular beam epitaxy (MBE) and low-pressure chemical vapor deposition (LPCVD), it has become feasible to make very thin compound semiconductor layers. By altering the composition of the elements, it is possible to fabricate layers with different band gaps. For example, by adding 20% aluminum to gallium arsenide, the band gap can be increased from 1.4 eV to approximately 1.6 eV. Because of the difference in lattice constants, a SiGe layer on a silicon substrate will be "strained" (the lattice spacing will be neither that of silicon or germanium). If SiGe is grown with 20% germanium to form a thin strained SiGe layer, the band gap is reduced from 1.1 eV to approximately 0.95 eV. This SiGe layer must be kept less than a critical thickness (a function of the Ge fraction) to maintain the required properties, but fortunately the conditions are compatible with the requirements for good BJT design.

This property can be used to advantage in the design of bipolar transistors if the larger-band-gap region is used for the emitter and the smaller-band-gap region is used for the base. In Chapter 4 we saw that the current gain I_C/I_B is given by Eq. (4.15):

$$\beta = (D_n n_{b0} W_e)/(D_p p_{ne0} W_b) \tag{7.35}$$

If we now use the mass-action law to write the expressions for minority-carrier thermal equilibrium concentrations using Eqs. (2.4) and (2.5) with a band gap E_{ge} in the emitter and E_{gb} in the base, we obtain:

$$n_{b0} = n_{ib}^2/N_{AB}$$
$$= N_c N_v \exp(-E_{gb}/kT) \tag{7.36}$$
$$p_{ne0} = n_{ie}^2/N_{DE}$$
$$= N_c N_v \exp(-E_{ge}/kT) \tag{7.37}$$

The expression for current gain thus becomes:

$$\beta = (D_n N_{DE} W_e)/(D_p N_{AB} W_b)\ \exp[E_{ge} - E_{gb})/kT] \tag{7.38}$$

If the band-gap difference between emitter and base is 0.15 eV, the gain is increased by a factor of approximately 400. Since a current gain of order 100 is sufficient for most circuit applications, this means that the base doping can be increased by a large factor, thus reducing the base resistance r_{bb}. The base resistance is a key design parameter for microwave bipolar transistors, since together with f_t and C_{jc} it determines the maximum oscillation frequency (4.79) and hence the high-frequency power gain.

The two main contenders for HBT structures are Si–SiGe–Si (which can be fabricated in integrated circuit technology) and GaAlAs–GaAs–GaAs (for microwave applications). Equation (7.24) gives the band-gap increase for GaAlAs compared to GaAs. Figure 7.29 shows a widely used impurity profile.

The base doping level is kept high to reduce r_{bb}. The emitter doping level is high at the surface to ensure good ohmic contact and to reduce minority carrier charge. The emitter doping N_2 is low (of order 10^{15} to 10^{16} cm^{-3}) to give a low value of junction capacitance (C_{je}). Also to minimize C_{jc}, the collector doping level is kept at a comparable value N_2 to a depth sufficient to maintain the required breakdown voltage, before the highly doped region is reached (for low collector resistance).

Graded base band gap HBT

For the SiGe HBT, an additional performance enhancement may be obtained if the germanium fraction is graded across the base region thickness W_b. Since a change in band gap ΔE over a distance W_b introduces a change in potential ΔV, an electric field $\Delta V / W_b$ is created. For 0.1 eV band-gap difference across a 0.1 μm base, this field will be 10^5 V/cm and can be in a direction such that the carriers are accelerated across the base with a drift current adding to the normal diffusion current. The transit time can be reduced significantly and hence the f_t increased. (See also CD-ROM BIPGRAPH examples SIGEA, SIGEB, SIGEC; "Effective doping vs depth" [to see effective doping gradient in the base region created by the Ge fraction gradient] and "Neutral base minority carrier conc. vs depth" [to see $n(x)$ in the base region for the 3 cases]. The base transit time TBASE can be seen in the .1st files [−1.0 ps for SIGEA, 0.47 ps for SIGEB, 7.8 ps for SIGEC.)

GaAs versus Si

GaAs has the advantage of a high electron mobility for *NPN* transistors (approximately 5000 cm^2/V s at moderate doping levels). This can give substantially lower base transit times than silicon (even with the graded base band gap of SiGe). However, the SiGe HBT has the advantage

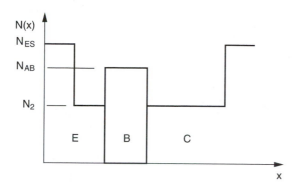

Figure 7.29 Impurity profile for a heterojunction bipolar transistor.

of being compatible with normal silicon processing technology (with the extra processing step for the SiGe layer). Since 1995 VLSI chips have been reported with SiGe devices incorporated, with bandwidth or speed advantages of approximately a factor of 2 over conventional silicon.

7.7 MOS MEMORY DEVICES

We shall not discuss here memories based on circuit techniques. The static random access memory (SRAM) is based on flip-flop circuits (both ECL-type bipolar and CMOS). The MOS transistor has a near-perfect capacitance at the input gate terminal. This may also be used as a storage element. In MOS dynamic memories (DRAMs), the charge is supplied and removed through the drain of a driving MOSFET. Because of the source–drain leakage current of the driving MOSFET, charge on the gate oxide of the memory element is lost quite rapidly. This arrangement is therefore only usable with "refreshing" circuitry and forms an essential part of most digital circuits.

A separate class of eraseable programmable nonvolatile memory elements (EPROMs) is also based on the MOS gate, but to avoid leakage of the charge, a floating gate (FAMOS) structure [7, 8] is used, as shown in Fig. 7.30.

The floating gate is made using doped polysilicon (see Chapter 8 for details) and is completely surrounded by SiO_2. The threshold voltage is such that normally the MOSFET is "on" and current will flow between source and drain if a potential V_{DS} is applied. If a negative

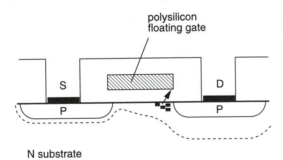

Figure 7.30 Cross section of FAMOS structure. After Frohman-Bentchkowski, [7].

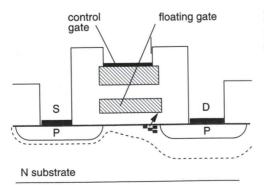

Figure 7.31 Cross section of electrically erasable stacked gate (SAMOS) structure. After Iizuka et al. [9].

charge is put on the gate, the device will be turned off. This may be achieved by applying a sufficiently high V_{DS} pulse so that avalanche injection occurs; the electrons at the drain end of the channel acquire sufficient energy to cross the insulating area (typically of order 0.1 μm thick) and build up a negative charge on the polysilicon gate. This charge will normally remain for many years and may be considered permanent, hence the name "nonvolatile memory." The device is called FAMOS for floating gate avalanche injection MOS. In its practical EPROM form, an array of such devices is made on one chip. Writing the memory states is done by suitable wiring of the array to apply the pulses. Erasure is possible using ultraviolet light of sufficiently high energy to let the electrons to pass over the insulating barrier back to the silicon channel region.

A modified version of the FAMOS structure [9] uses two gates and is electrically eraseable. This is the EEPROM and is shown in Fig. 7.31. The additional gate is used as a control element. The writing operation occurs as in the FAMOS device, with a high-voltage between drain and source. The electron charge on the floating gate can be removed by applying a high positive voltage pulse to the control gate, with substrate, source, and drain at zero volts. The electrons acquire enough energy to pass through the top oxide to the top gate and the structure is then ready for reprogramming.

7.8 CONCLUSIONS

We have attempted to introduce the reader to a wide range of devices outside the conventional integrated circuit VLSI BJT and MOSFET structures. The treatment has necessarily been superficial but, it is hoped, adequate to understand the essential physical phenomena and to provide in some cases the basic relationships between the device structure and the electrical characteristics.

PROBLEMS

1. Show by integrating Eq. (7.4) for a photodiode that the maximum efficiency obtainable due to photons absorbed within the depletion-layer thickness w_{scl} is given by Eq. (7.13)

$$\eta = [1 - \exp(-\alpha w_{scl})].$$

2. Consider a silicon photodiode with incident light at a wavelength of 0.85 μm. If the incident photon flux density is 10^{12} cm^{-2} and the absorption coefficient is 900 cm^{-1}, calculate (a) the depletion-layer thickness necessary to absorb 80% of the incident light (assume a junction close to the surface); (b) the applied reverse bias necessary if the background doping level is 10^{15} cm^{-3}; and (c) the current that will flow if the diode has a diameter of 100 μm. Calculate the approximate signal bandwidth of the above diode neglecting any circuit effects.

3. A P^+NN^+ photodiode doped $N_D = 10^{15}$ cm^{-3} over a thickness of 10 μm with a diameter of 100 μm is used to detect 0.85 μm wavelength light with an absorption coefficient of 900 cm^{-1}; the carrier lifetime is 100 ns. Calculate: (a) the minimum reverse bias required for maximum efficiency; (b) the 3 dB cutoff frequency; (c) the dark current at the specified minimum bias; (d) the minimum detectable light power for a signal-to-noise ratio of 10 dB and a bandwidth of 100 MHz (neglect amplifier noise and make suitable approximations).

4. A solar cell 1 cm square is made with a surface-diffused layer of sheet resistance 50 ohm/square. There are 5 narrow metal contact fingers, spaced uniformly and connected at the end by a 1 cm-wide

stripe. Estimate the series resistance. If the cell has an efficiency of 15% without the resistance, a dark current $I_0 = 1$ nA and the incident sun power is 50 mW, calculate, making suitable approximations: (a) the short-circuit current I_{ph}; (b) the open-circuit voltage; (c) the power loss due to the resistance when used with a load giving a current of $I_{ph}/2$.

5. Calculate the position of the Fermi level for a laser PN diode doped 2×10^{20} cm^{-3} on each side. Plot the Fermi probability distribution around E_F on the P side and hence argue the existence of a partially empty region between E_F and E_V as in Fig. 6.10(a).

6. A CCD cell measures 2×2 μm^2; the MOS oxide thickness $t_{ox} = 0.025$ μm; the doping level is 10^{15} cm^{-3} and the flat-band voltage is 1 V. If a gate voltage of 5 V is applied, calculate: (a) the initial value of input capacitance; (b) the final value of input capacitance at low frequency; (c) the final high-frequency value of input capacitance.

7. An interdigitated bipolar transistor has a maximum current density limit of 10^4 A/cm^2 due to the Kirk effect. The lithography places a lower limit of $L = 2$ μm on the emitter stripe diffusion. If the maximum stripe length-to-width ratio B_s/L is 50, calculate the total emitter area and number of stripes required to allow a maximum collector current of 500 mA. If the active base sheet resistance ρ/W_b is 5000 ohm/square, calculate the final value of base resistance r_{bb}.

8. A heterojunction BJT is made using an SiGe layer for the base. If the current gain of a similar transistor with a silicon base is 5, what band-gap reduction is required in the base in order to obtain a gain of 100? If a Ge fraction of 20% is used in the base, by what factor can the base doping be increased to maintain the required gain of 100? What improvement in maximum oscillation frequency will this give compared to the Si base transistor?

9. A Darlington BJT pair is to be designed for which the current density limit is $J_K = 1600$ A/cm^2, the current gain is $\beta = 20$, the minimum emitter width $L = 5$ μm, and the maximum emitter stripe length is $B_S = 100$ μm. Calculate: (a) the emitter areas and total stripe lengths for each transistor for a maximum total collector current of 500 mA; (b) the number of stripes for each transistor.

10. Calculate the varactor diode area and cut-off frequency at one-quarter the breakdown voltage, if the breakdown voltage is to be 20 V and the minimum capacitance at breakdown is to be 0.1 pF.

11. An IMPATT diode can be made without the narrow N^+ region using the peak electric field to create the necessary impact ionization. Consider the diode considered in the example of Section 7.6.3 but without the narrow N^+ region, with the same peak electric field, and with a uniform doping level of 1.0×10^{16} cm^{-3}. Draw the electric field versus distance for both diodes and note the distance over which the field exceeds 3×10^5 V/cm (this gives a rough estimation of the degree of impact ionization). Calculate the total voltage.

12. Using Eq. (7.16), show that for large incident photocurrent, the signal-to-noise ratio tends to $S/N = \eta W_{0pt}/4h\nu B$. This is the quantum-noise-limited value.

References

1. S. M. Sze, *Physics of Semiconductor Devices*, 2nd ed. New York: John Wiley & Sons Inc., 1981.
2. E. S. Yang, *Microelectronic Devices*. New York: McGraw–Hill, 1988.
3. D. J. Roulston, *Bipolar Semiconductor Devices*. New York: McGraw–Hill, 1990.
4. J. Baliga, *Modern Power Devices*. New York: John Wiley & Sons, 1987.
5. M. S. Tyagi, *Introduction to Semiconductor Materials and Devices*. New York: Wiley, 1991.
6. D. A. Neamen, *Semiconductor Physics and Devices*, 2nd Ed. Chicago: Irwin, 1997.

7. D. Frohman-Bentchkowsky, "FAMOS—A New Semiconductor Charge Storage Device," *Solid State Electronics* **17**, 517 (1974).

8. J. J. Chang, "Nonvolatile Semiconductor Memory Device," *Proc. IEEE* **64**, 1039 (1976).

9. H. Iizuka, F. Masuoka, T. Sato, and M. Ishikawa, "Electrically Alterable Avalanche Injection Type MOS Read-Only Memory with Stacked Gate Structures," *IEEE Trans. Electron Devices* **ED-23**, 379 (1976).

10. A. Bar-lev, *Semiconductor and Electronic Devices*, 3rd Ed. New York: Prentice-Hall, 1993.

11. Y. Tarui, Y. Hayashi, and T. Sekigawa, "Diffusion Self-Aligned Enhance-Depletion MOS-IC," *Proc. 2nd Conf. Solid State Devices Suppl. J. Jpn. Soc. Appl. Phys.* **40**, 193 (1971).

12. F. E. Holmes, C. A. T. Salama, "VMOS—A New MOS Integrated Circuit Technology, *Solid State Electronics* **17**, 791 (1974).

Chapter 8

Silicon Chip Technology and Fabrication Techniques

8.1 INTRODUCTION

In this chapter we provide an overview of fabrication methods for silicon chips. The starting point is the crystalline silicon wafer, which is often provided in the form of a substrate and a surface layer of specified doping type and resistivity. Oxide growth is an essential step and is covered next, together with lithographic techniques. The introduction of donor and acceptor impurity atoms in a controlled manner is crucial to the device fabrication and the two widely used methods, solid-state diffusion and ion implantation, are described; this is followed by a very brief overview of metallization and interconnect stages.

The basic processes used to fabricate diodes, BJTs, and MOSFETs are described. A brief description of some additional fabrication steps (polysilicon deposition, oxide isolation, trench isolation) used in state of the art industrial devices is also provided. The chapter concludes with a summary description of current frequently used technologies for bipolar, CMOS, and BiCMOS structures. This description includes cross-section and mask layout diagrams to highlight the essential features of each technology.

8.2 WAFER FORMATION

8.2.1 Crystal growth

Wafer preparation consists of several steps, starting with the raw material—sand—and ending with the completed silicon wafer ready for processing. The sequence for wafer formation is shown in Fig. 8.1.

The sand (SiO_2) is heated in a furnace in the presence of carbon compounds. As a result, silicon with about 98% purity is obtained. It may be noted that this represents an impurity fraction of 2 parts per 100. We recall that for a typical donor or acceptor impurity concentration of $10^{15} cm^{-3}$, the impurity fraction is about 1 part per 10^8. Clearly a very high degree of purity is required in order to make silicon devices. Pulverization of this silicon and

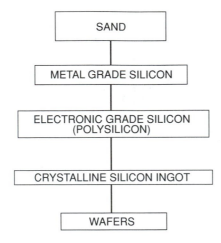

Figure 8.1 Flow chart for formation of silicon wafers.

treatment with hydrogen chloride (HCl) produces trichlorosilane ($SiHCl_3$), which is liquid at room temperatures. Unwanted impurities are removed by fractional distillation, and the purified $SiHCl_3$ is then used in a hydrogen reduction reaction to provide electronic-grade silicon (EGS). This is polycrystalline silicon of very high purity and is the starting point for crystal growth.

Silicon crystal growth is performed using either the Czochralski technique or the float-zone process. We will discuss here the Czochralski technique, widely used for silicon, in which a crystal rod is pulled, starting from a seed crystal, from the melt, as shown in Fig. 8.2. The polycrystalline silicon is placed in the crucible and heated above the melting point of silicon. A seed crystal is used to initiate the process by inserting it into the melt. It is then slowly raised (as the holder is rotated). At the solid–liquid interface, freezing occurs, producing single crystal silicon.

Dopant is added to the melt to create a crystal of specified impurity type and resistivity. After the crystal ingot is grown, the wafers must be formed. First a flat region is ground along the ingot to denote the crystal orientation. The ingot is then sliced into wafers using a diamond cutting saw [we recall that diamond (carbon) is mechanically much harder than silicon]. The wafers are then lapped to produce a uniform flatness; this is followed by etching to clean the surface and by polishing.

8.2.2 Epitaxial layer growth

In many devices, a surface layer of silicon of specified doping, different in dopant type and in resistivity form the bulk wafer, is required. This layer is made using the epitaxial technique and the layer is referred to as an *epitaxial layer*. Vapor-phase epitaxy is normally used for silicon. The substrate wafer acts as a seed crystal, and the layer is grown at a temperature well below the melting point. A common source for the growth is silicon tetrachloride ($SiCl_4$), giving the reaction:

$$SiCl_4(gas) + 2H_2(gas) \Longrightarrow Si(solid) + 4HCl(gas) \tag{8.1}$$

The dopant (e.g., boron, antimony) is introduced at the same time as the silicon tetrachloride in a contolled amount to provide the required resistivity. The epitaxial layer thickness is typically

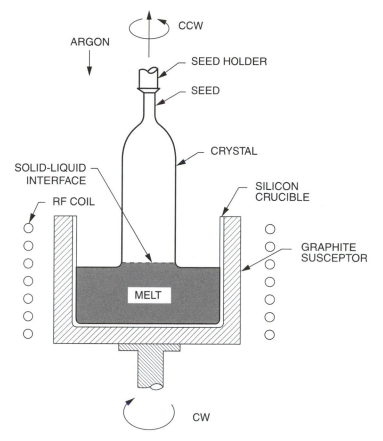

Figure 8.2 Czochralski crystal puller. After S. M. Sze [5]. Reprinted by permission of John Wiley & Sons, Inc.

in the range 2 to 8 μm and has the same crystal orientation as the substrate, although the doping can be of opposite type for bipolar integrated circuit chips (the same type for discrete BJTs).

8.3 OXIDE LAYERS AND LITHOGRAPHY

8.3.1 Oxide growth

If silicon is allowed to come in contact with oxygen, a thin surface layer of silicon dioxide, SiO_2, is formed. Growth of this layer is accelerated at high temperatures and in the presence of steam (this is similar to what happens to iron, when oxide layers grow in the form of rust). This type of layer is referred to as *thermal oxide* and is used to "isolate" the silicon surface. By removing the oxide layer in selected areas, dopant atoms can be introduced in a controlled manner, to be discussed under "lithography." Thin oxide layers (of order 100 nm or less) are also an essential part of MOS devices between the gate and the channel.

The chemical reaction for oxide growth is of one of two forms:

$$Si + O_2 \Longrightarrow SiO_2 \tag{8.2}$$

$$Si + 2H_2O \Longrightarrow SiO_2 + 2H_2 \tag{8.3}$$

It is important to note that as the SiO_2 layer is created, silicon atoms are "used up," and the oxide layer ends up partly below and partly above the original silicon surface. In fact, 44% of the oxide layer is below the original silicon surface. Table 8.1 shows the rate of growth of the SiO_2 layer as oxidation time and temperature are varied.

Results are given for dry and wet growth conditions, the latter being obtained by using a steam ambient. By comparing the thickness after a time of 1 h, it is clear that a wet ambient accelerates the growth by a factor of about five. The actual choice of temperature and time are dictated by the fact that if the temperature is too high, the time is too short, and accurate control of the layer thickness is not possible. On the other hand, very long times tie up the equipment and add to the manufacturing costs. Temperatures around 1000°C and times up to 1 h are typical. For MOS structures, the gate oxide is usually grown under dry conditions to give better control and high-quality layers. Modern MOS devices have gate oxides as thin as 30 nm (0.03 μm, 300 Å).

8.3.2 Lithography

Photolithography is widely used to define areas on the surface of the wafer through which dopant atoms are introduced to create P and N layers of diodes, BJTs, and MOSFETs. It is also used to define the final metal pattern on the wafer.

TABLE 8.1A
Rates of oxide growth (μm) versus temperature and time for dry oxidation for 100 orientation. Data extracted from Meindl et al. [8].

Temperature (°C) Time (h)	800	900	1000	1200
0.3	—	—	0.023	0.094
1.0	—	0.018	0.050	0.195
3.0	0.013	0.038	0.10	0.35
10.0	0.03	0.087	0.23	0.68
30.0	0.065	0.19	0.40	1.1

TABLE 8.1B
Rates of oxide growth (μm) versus temperature and time for wet oxidation for 100 orientation. Data extracted from Meindl et al. [8].

Temperature (°C) Time (h)	900	1050	1250
0.3	—	0.27	0.54
1.0	0.23	0.55	1.0
3.0	0.45	0.95	1.8
10.0	0.93	1.8	3.5

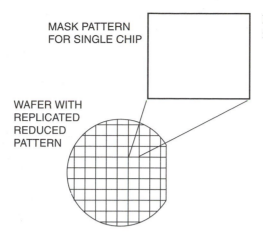

MASK PATTERN
FOR SINGLE CHIP

WAFER WITH
REPLICATED
REDUCED
PATTERN

Figure 8.3 Replication and reduction of an individual circuit mask pattern onto a master mask.

The first step is to make a mask pattern that describes the geometry of the regions into which impurities are to be introduced. The mask is made using standard photographic techniques, consisting of a large-scale (computer-generated) pattern, reduced optically to the required size. Simultaneously the pattern is usually stepped and repeated to form a matrix of identical patterns replicated over the entire wafer area. This procedure is illustrated in Fig. 8.3. Direct-write techniques using electron beam methods are also used. After all processing is completed, the wafer will be sliced and each individual square of the replicated pattern will form the silicon die, to be mounted in a header and wire bonded to the pins.

The surface of the wafer, with an oxide layer of order 1 μm thickness already prepared, is coated, by spinning, with photoresist, a material sensitive to the ultraviolet light used for exposure. The photoresist is not sensitive to light of wavelength greater than about 0.5 μm; so yellow light can be used, hence the name "yellow" room.

Baking is used to remove the solvent and to give good adhesion. The mask is interposed between the uv light source and the coated wafer, as shown in Fig. 8.4. In the case of positive resist, the exposed regions become soluble and are removed during development. For negative resist the reverse occurs. This is followed by rinsing and drying.

The exposed oxide layer is then removed by etching, using, for example, buffered hydrofluoric acid (which attacks SiO_2 but not the resist). Figure 8.5(a)–(c) shows these final steps. The wafer is now ready for the introduction of donor or acceptor impurities.

Limits of optical lithography

As smaller and higher-speed devices are required, finer linewidths are necessary, since the definition (minimum resolution) is a direct function of the wavelength of light used. At present (1997) 0.35 μm technology is in production. Deep ultraviolet technology uses 248 nm light; 193 nm laser sources are being studied for the next generation of integrated circuits. The reduction to finer and finer linewidths necessitates development of suitable photoresists combined with lenses that will perform correctly at these very short wavelengths. An excellent overview is given in Ref. [9].

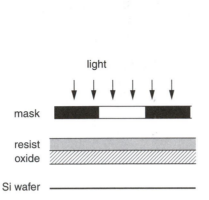

Figure 8.4 Exposure of mask pattern on to wafer with SiO₂ layer coated with photoresist.

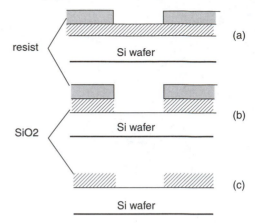

Figure 8.5 (a) Photoresist opening after developing and washing; (b) oxide removal to form window; (c) removal of remaining resist.

8.4 INTRODUCTION OF DOPANT ATOMS

There are two primary methods used for introducing dopant atoms into the wafer: solid-state diffusion and ion implantation. In the former case, a gas at high temperature containing the dopant species is in contact with the silicon surface in the regions determined by the windows. Because of the high temperature, the dopant atoms diffuse (slowly) into the silicon. In the second method, a high-energy beam of ions containing the dopant species penetrates the surface of the silicon, thus embedding the dopant atoms below the surface. Figure 8.6 illustrates the two methods of impurity atom doping. We will now examine these two methods in order to understand their relative advantages.

8.4.1 Solid-state diffusion

Diffusion of impurity atoms takes place in a high-temperature furnace. The wafers are placed in a quartz tube, and gas containing the dopant atoms flows over the wafers (with windows opened in the oxide layer in the required locations). The dopant atoms (phosphorous, arsenic for *N*-type, boron for *P*-type dopants) diffuse into the silicon. The dopant atoms diffuse either substitutionally (replacing silicon atom sites in the lattice) or interstitially (between silicon atom sites).

The atoms diffuse into the semiconductor material, obeying (to a first approximation) Fick's laws of diffusion. The rate of flow F is proportional to the concentration gradient:

$$F = -D \frac{dC}{dx} \tag{8.4}$$

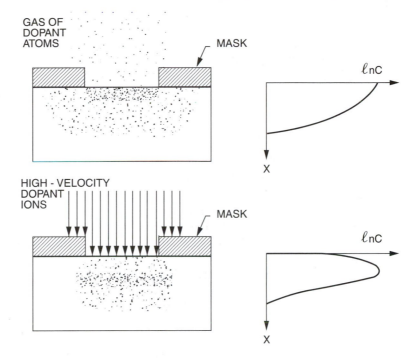

Figure 8.6 Doping the silicon: top diagram, solid-state diffusion; lower diagram, ion implantation. After S. M. Sze, [5]. Reprinted by permission of John Wiley & Sons, Inc.

where D is the atomic coefficient of diffusivity and C is the dopant concentration at a depth x.

This leads to the time-dependent form of the diffusion equation:

$$\frac{dC}{dt} = D\frac{d^2C}{dx^2} \tag{8.5}$$

The value of D is strongly temperature dependent (clearly at room temperature, practically no atoms will diffuse significantly in a period of 1 year, but at a temperature of 1000°C, diffusion to a depth of several microns will occur in a period of 1 h).

$$D = D_0 \exp(-E_a/kT) \tag{8.6}$$

where D_0 is the value of D extrapolated to $T = \infty$ and E_a is the activation energy and is in the range 0.5 to 5 eV depending on the diffusion mechanism. Table 8.2 gives the nominal values of D_0 and E_A for some commonly used dopants.

Figure 8.7 shows D as a function of temperature for three commonly used dopants in silicon. Note the slow rate of diffusion of arsenic and the intermediate values for boron and phosphorous. Gallium (Ga) and aluminum (Al) diffuse even faster than boron. The choice of dopant is important for both electronic and economic reasons. Regions of the device that are

TABLE 8.2
Values of D_0 and E_a for several dopants

	Phosphorus (P)	Boron (B)	Arsenic (As)
D_0 (cm²/s)	10.5	10.5	0.32
E_a (eV)	3.69	3.69	3.56
Temp. range (°C)	950–1235	950–1275	1095–1380

doped early in the process require a choice of dopant atoms that will not alter substantially their distribution during subsequent processing, As and Sb are obvious choices and are used in the buried layer of bipolar IC chips (see Section 8.6.2). Gallium and aluminum are used in very deep-junction high-voltage structures, to avoid excessively long periods in the high-temperature furnace. For a P-type dopant in integrated circuit applications, boron is universally used. The choice of phosphorus or arsenic for the N-type dopant depends on several factors and both are in fact widely used.

It is important to note that diffusion through an SiO_2 layer occurs at a much slower rate. The diffusivity at 900°C in silicon dioxide is 4×10^{-19} cm²/s for arsenic, 3×10^{-19} cm²/s for boron and 10^{-18} cm²/s for phosphorus [5]. This means that an oxide layer of order 1 μm is sufficient to prevent these impurity atoms diffusing into the silicon.

The electrical properties of a semiconductor device depend critically on the junction depths and also on the shape of the impurity atom distribution—the impurity profile. There are two distinct mathematical forms for $C(x)$ depending on the method used to diffuse the atoms, (1) infinite- and (2) limited-source diffusion.

Infinite-source (constant surface concentration) diffusion

In this case the gas containing the dopant atoms remains in contact with the silicon surface throughout the duration of the high-temperature diffusion. The surface concentration C_s will therefore remain constant, at a value determined by the solid solubility of the dopant in the semiconductor. This value is of order 10^{21} cm⁻³ for most dopants. The atoms diffuse deeper below the surface with time as shown in Fig. 8.8. The distribution follows a complementary error function law:

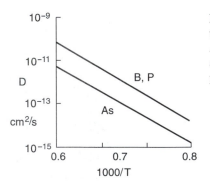

Figure 8.7 Diffusivity D (cm²/s) as a function of reciprocal temperature $1000/T$ (K) for three common dopants. Antimony (Sb) has a lower diffusivity than arsenic (As); aluminum (Al) has a higher diffusivity than boron or phosphorus.

$$C(x) = \text{erfc}\left(x/2\sqrt{Dt}\right) \tag{8.7}$$

This is the (ideal) result obtained for a conventional phosphorus diffusion when creating the emitter of a BJT, although significant deviations from the erfc law occur in practice.

Limited-source diffusion

The second basic type of diffusion occurs when the gas source is cut off after a certain time and the atoms allowed to continue diffusing into the silicon. This occurs when an initial diffusion of type (1) is made and then an oxide layer is grown in the window region to prevent further introduction of impurities. If the wafer is left in the high-temperature furnace, the impurity atoms will continue to diffuse into the semiconductor material. If we assume that the initial time is short compared to the time elapsing after the oxide layer is grown (for the same temperature), the final depth of the diffusion is much greater than the original (erfc function) distribution and the mathematical law is a Gaussian function:

$$C(x) = C'_s \exp(-x^2/4Dt) \tag{8.8}$$

where $C'_s = Q_0/\pi\sqrt{Dt}$ and Q_0 is the integral of the impurity atom distribution before the supply is cutoff, that is, Q_0 is the integral of Eq. (8.7) for an initial time t_1 and is given by:

$$Q_0 = 2C_s\sqrt{Dt/\pi} \tag{8.9}$$

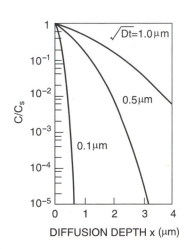

Figure 8.8 Impurity atom distribution for infinite-source diffusion. After S. M. Sze, [5]. Reprinted by permission of John Wiley & Sons, Inc.

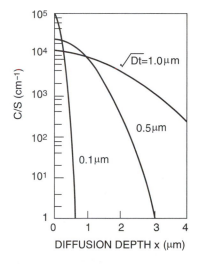

Figure 8.9 Gaussian distribution of impurity atoms for a limited-source diffusion. After S. M. Sze, [5]. Reprinted by permission of John Wiley & Sons, Inc.

Figure 8.9 shows the Gaussian impurity distribution for several values of normalized time. This is the resultant impurity profile for most base diffusions in bipolar transistors and also for most source and drain diffusions in CMOS technology. Since the Gaussian function is mathematically more tractable than the error function, it is frequently used as an approximation. Figure 8.10 shows the erfc and Gaussian distributions superimposed. Apart from the difference in gradient at $x = 0$, the two curves can be made to coincide over a very wide range. Furthermore, an excellent fit can be made to either distribution well below the surface $(x > 4\sqrt{Dt})$ by an exponential function, $C(x) \propto \exp(-x/x_0)$ (a straight line on a log–linear plot). In many cases the exponential approximation is adequate, especially when it is recognized that both the above erfc and Gaussian solutions are only approximations to the actual impurity distributions, due to anamolous diffusion created by interactions between different species of the diffusing atoms. The values of diffusivity given in Table 8.2 and the resulting curves of Fig. 8.7, for example, are only valid for doping levels below about 10^{18} cm^{-3}. For greater doping levels the diffusivities increase substantially [5].

Example:
   A bipolar transistor is made on an N-type epitaxial layer doped 10^{16} cm^{-3} with a predeposition of boron at 980°C for 5 min, followed by a drive-in for 80 min at the same temperature. The emitter is created by an infinite-source phosphorus diffusion at 980°C for 5 min.
   From Fig. 8.7 for boron at $980 + 273 = 1253$ K, $1000/T = 0.8$; $D = 10^{-14}$ cm^2/s.
   For 5 min (300 s) $\sqrt{Dt} = 1.7 \times 10^{-6} = 0.017 \,\mu$m.
   From Eq. (8.9), assuming a solid solubility limit for phosphorus $C_s = 10^{21}$ cm^{-3}, we find
$Q_0 = 2\sqrt{(Dt/\pi)}C_s = 0.02 \times 10^{-4} \times 10^{21} = 2 \times 10^{15}$ cm^{-2}.
   For the boron drive-in phase, $t = 85 \times 60$ s and hence $\sqrt{Dt_2} = 4 \times 0.017 \,\mu$m $= 0.068 \,\mu$m.
   The surface concentration C_{s2} at the end of the drive-in, from Eq. (8.8) is:
$Q_0/\pi\sqrt{Dt_2} = 2 \times 10^{15}/(\pi \times 0.068 \times 10^{-4}) = 2 \times 10^{19}$ cm^{-3}.
   The collector–base junction depth can be estimated from Fig. 8.9 when the boron doping level has fallen to the epitaxial layer doping of 10^{16} cm^{-3} that is, a ratio of 1/2000; the corresponding depth is approximately $6\sqrt{Dt_2} = 6 \times 0.07 = 0.42 \,\mu$m.
   The emitter–base junction depth occurs when the phosphorus and boron concentrations are equal. This occurs at a doping level of approximately 10^{19} cm^{-2} (the boron impurity profile has only fallen slightly), a reduction ratio of approximately 0.01 and Fig. 8.8 enables us to estimate the junction depth as about $3 \times \sqrt{Dt_1} = 3 \times 0.017 \,\mu$m $= 0.05 \,\mu$m.
   More accurate results may be obtained using the tabulated values of D_0 and E_A, and by solving the erfc and Gaussian functions.

Sideways diffusion

An important aspect of solid-state diffusion in the fabrication of silicon devices is the fact that the impurity atoms diffuse sideways as well as vertically into the wafer. Figure 8.11 shows this effect once the diffusion is complete. As a rule of thumb, the lateral diffusion of dopant atoms

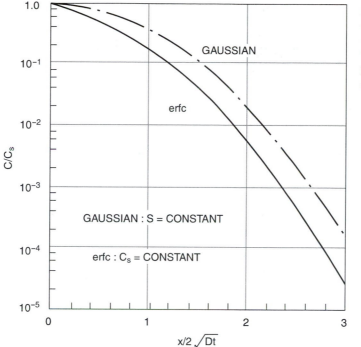

Figure 8.10 Gaussian and erfc impurity distributions superimposed. After S. M. Sze, [5]. Reprinted by permission of John Wiley & Sons, Inc.

can be approximated by saying that the lateral location of the junction (where the diffused profile equals the constant background doping level) is 0.8 times the vertical junction depth. This is most important when determining the minimum design rules for mask layout, since the diffused area of a layer will always be greater than the mask (window) dimensions. It is also important when estimating the sidewall capacitance of a junction. This can usefully be approximated by considering an arc equal to that of a quarter cylinder of radius $0.8X_j$, where X_j is the junction depth.

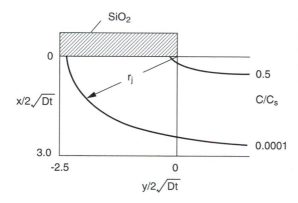

Figure 8.11 Impurity concentration contours at the edge of an oxide window, demonstrating sideways diffusion. The diffusion mask is the oxide covered area. After D. P. Kennedy and R. R. O'Brien, [7].

8.4.2 Ion implantation

For ion implantation, a high-energy (in the range 20 to 200 keV) beam containing ions with the dopant species is incident on the wafer surface. In the areas created by the window openings (as determined by the mask pattern), the dopant atoms penetrate below the surface. The higher the energy, the deeper the penetration. The impurity atom distribution is approximately Gaussian, with a peak value at a depth (below the silicon surface) called the projected range R_p:

$$C(x) = \left[S/\sqrt{2\pi}\, X_s \right] \exp\left[-(x - R_p)^2/2X_s^2 \right] \tag{8.10}$$

where X_s is referred to as the straggle, $\triangle R_p$. S is the called the dose of the implant and depends on the total duration of the high-energy implant.

Both projected range and straggle are directly related to the energy on the ion implant. Tables 8.3(a) and (b) show the values for typical dopants and energy values.

For the energy range 10 to 200 keV, it is clear that quite accurate implants can be made to depths less than 1 μm. Ion implants are thus used primarily for accurate control of the impurity distribution in shallow-junction devices, that is, in high-speed or high-frequency applications. Note that the projected range in an oxide layer is comparable to or less than that for silicon. SiO_2 layers of about 1 μm thick or less can therefore be used to prevent significant penetration of dopant atoms in much the same way that they are used as a diffusion barrier.

Because of the nature of the implant, in which many collisions occur between the high-energy atoms and the silicon lattice atoms, a trail of damage to the crystal lattice is created in the path of the implanted atoms, with most of the damage being located between 0 and R_p. This damage is detrimental to obtaining good electrical characteristics (in particular, high recombination rates go hand in hand with damage to the crystal). A necessary step following ion implantation is the annealing of the wafer to remove the damage and re-form the crystal structure. This is a short thermal step. This is carried out at temperatures in the range 700–900°C for times of order 30 min, although higher temperatures for very short times (of order seconds) are now also used (rapid thermal annealing, RTA).

TABLE 8.3A
Projected range R_p in μm for boron, phosphorus, and arsenic in silicon at various energies. After K. A. Pickar, [10].

Energy (keV)	10	30	100	300
Boron	0.03	0.1	0.31	0.73
Phosphorus	0.015	0.04	0.13	0.40
Arsenic	0.01	0.023	0.06	0.23

TABLE 8.3B
Ion straggles ΔR_p in μm for boron, phosphorus, and arsenic. After K. A. Pickar, [10].

Energy (keV)	10	30	100	300
Boron	0.015	0.034	0.07	0.10
Phosphorus	0.007	0.018	0.045	0.10
Arsenic	0.0037	0.008	0.023	0.055

A frequent approach for the formation of shallow junctions is to use ion implantation to set the initial volume of impurity atoms, and then to follow this with a thermal diffusion step (which also anneals out the damaged region). This is widely used for example in advanced BJTs to create the base region. Ion implantation gives much more accurate control of the total number of impurity atoms than does a short infinite-source diffusion step.

8.5 METALLIZATION AND INTERCONNECT STEPS

The final steps in wafer processing consist in contact window opening, metal deposition, and etching to create the metal interconnect pattern.

A photolithographic step is used with a "contact opening" mask to expose all regions of the silicon where contacts are required (e.g., source, drain, and gate regions of a MOS structure). Minimum design rule dimensions are normally used, which today means windows of about 1 μm square or less. Diagrams for various devices are given in Section 8.6. A metal, frequently aluminum (with a conveniently low resistivity of 2.7 μohm cm and good adherence properties to SiO_2), is then evaporated or sputtered onto the whole wafer. This is performed in vacuum and readily allows a metal film thickness of order 1 μm.

The metallized surface of the wafer is now treated to a final lithographic process, to remove the metal in all areas not required for interconnect and bonding pad formation. The bonding pads are relatively large square areas of metal, usually around the perimeter of the die, to which wires are bonded for connection to the header pins.

8.6 FABRICATION STEPS FOR TYPICAL SILICON DEVICES

Here we give an overview of the major steps involved in the fabrication of some common silicon devices. The intent is not to go into detailed descriptions of the technology, but rather to give the reader an overview of the important steps involved and of the differences between different technologies. In most cases, we give cross-section diagrams showing the various N- and P-type layers, together with mask layout diagrams.

8.6.1 Diodes

8.6.1.1 Discrete high-power diode

The diode is, of course, the simplest device to fabricate. In its simplest form, as used in high-power high-voltage structures, no mask steps are involved. The whole wafer area is used for a single diode, in order to provide very high-current rectifiers. The starting wafer is lightly doped (in order to provide high breakdown voltages), typically of order 10^{14} cm^{-3} and may be N or P-type. Figure 8.12 shows the cross section and the impurity atom distribution (usually referred to simply as the impurity profile) of the completed device.

If the background doping is N type, a first N^+ diffusion (phosphorus is a common dopant) is made to the back side of the wafer; this provides a good ohmic contact with low resistance and can also in some cases be used to improve the dc characteristics because of the gradient

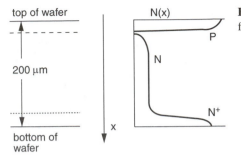

Figure 8.12 Cross section of wafer and impurity profile for a high-voltage diode.

introduced by the NN^+ doping profile. The final step is a P^+ diffusion from the top surface. This may be boron, but for deep junctions (required for high breakdown voltage) gallium or aluminum is also used. Both top and bottom surfaces are the metallized, and the diode is ready for packaging.

8.6.1.2 Planar diode

The planar diode is used in integrated circuit applications and also in some microwave devices. It is usually a low- or medium-voltage device, the starting wafer being an N_{epi} epitaxial layer on an N^+ substrate. Only a single P diffusion is required. This is made by opening up a window in the surface oxide layer, as described in Section 8.3. Figure 8.13 shows the cross section and mask diagrams.

Figure 8.13(a) shows the single P diffusion and the corresponding mask. Figure 8.13(b) shows the contact window opening, with a smaller mask; in Figure 8.13(c) the metal has been deposited and etched to form the interconnect to another part of the circuit (or to the bonding pads). It is important to understand this simple sequence, since it serves as the basis for more

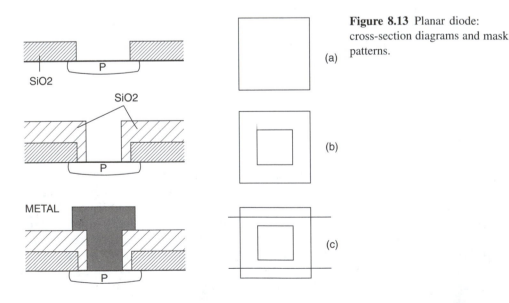

Figure 8.13 Planar diode: cross-section diagrams and mask patterns.

complex devices in the following examples. Typical dimensions for a small low-voltage diode are of order 3 μm square for the contact opening and 6 μm square for the P diffusion window. Note the sideways (lateral) diffusion, discussed in Section 8.4.1. Some special diode types were discussed in Chapter 7.

8.6.2 The conventional *NPN* integrated circuit bipolar transistor

This, together with the MOSFET, is the most important component in integrated circuit technology. Here we examine the conventional (older) BJT technology using a metal-contacted emitter and P^+ isolation. In Section 8.8 we look at a modern BJT with a polysilicon emitter and oxide plus dielectric (trench) isolation. Figure 8.14 shows the essential steps in the fabrication of an integrated circuit BJT. The wafer is lightly doped P type, typically $N_a = 10^{15}$ cm^{-3}. The first step is the creation of a heavily doped (typically 10^{18} cm^{-3}) N^+ buried layer. This will exist in every part of the chip where a transistor is to be located (and is also normally used in areas where P-diffused resistors are to be made). The main purpose of this N^+ buried layer is to provide a low-resistance path to current flowing horizontally to the collector contact. The second processing step consists in depositing an epitaxial layer of order 2 to 8 μm thickness. This will form the collector region of the BJT and the epitaxial doping level N_{epi} is chosen for the required breakdown voltage, typically 10^{16} cm^{-3} for IC structures, where the supply voltage is of order 5 V.

Once the epitaxial layer has been deposited, the second mask step is applied. Figure 8.14(b) shows the introduction of a deep P^+ isolation diffusion. This must penetrate down through the epitaxial layer to provide a continuous P region to the P substrate. The purpose of this step is to completely surround a single transistor with a P region, so that between one transistor and

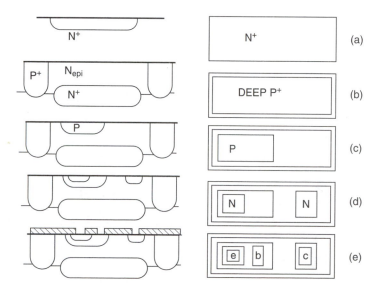

Figure 8.14 Integrated circuit *NPN* BJT fabrication steps: (a) buried layer; (b) deep P^+ isolation diffusion; (c) base diffusion; (d) emitter and collector N diffusions; (e) contact windows.

another adjacent one on the chip, there is always an *N–P–N* path. The *P* isolation is biased at the most negative potential in the circuit, so that the isolation junctions are always reverse biased; this provides dc isolation and maintains a low depletion-layer capacitance, desirable for good high-speed performance.

Figure 8.14(c) shows the formation of the *P* base diffusion. This is similar to the diode step discussed above.

Figure 8.14(d) shows the formation of a diffused *N* layer for the emitter. Simultaneously an *N* diffusion is created for the collector contact. In high-performance BJTs, the collector diffusion is performed immediately after the *P*+ isolation diffusion and uses a longer diffusion time so that the *N*+ collector extends down to the buried layer. This gives a lower collector series resistance; this is at the expense of an extra mask and increased processing time.

Figure 8.14(e) shows the formation of contact windows. Metallization patterns are not shown. These will depend on the circuit configuration, but in all cases the final metal pattern must extend past the contact window openings, so that no surface silicon is in contact with the atmosphere.

Minimum mask dimensions for window openings are about 2 μm for older fabrication facilities to 0.5 μm for state of the art facilities. The corresponding performance figures are usually quoted in terms of the f_t of the transistor, which can be anywhere from 1 GHz for the older technologies with emitter–base junction depths of order 0.5 μm and base–collector junction depths of order 1.0 μm to 25 GHz for current advanced structures with corresponding junction depths of order 0.1 and 0.2 μm, respectively.

Figure 8.15 shows the mask diagrams aligned with the cross-section diagrams and the impurity profile aligned with the cross section. The critical region of the collector is the residual

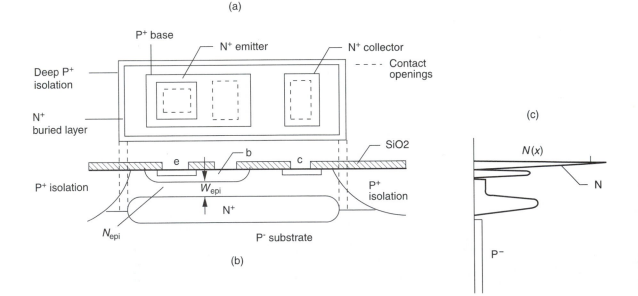

Figure 8.15 Integrated BJT showing: (a) superimposed mask layout; (b) cross section aligned with (a); (c) impurity profile aligned with (b). Reproduced with permission from: D. J. Roulston, *Bipolar Semiconductor Devices*, McGraw–Hill, New York, 1990.

epitaxial layer thickness marked W_{epi}. This, together with the doping level N_{epi}, determines the breakdown voltage of the collector–base junction, and hence the maximum dc operating voltage.

8.6.3 The metal gate MOSFET

Figure 8.16 shows the main steps involved in making a metal gate MOSFET. The source and drain are diffused conventionally. The thin gate oxide region is defined with a metal pattern that overlaps the source and drain diffusions to ensure inversion of the channel region when bias is applied (if no overlap occurs, with a gap at either end, the inversion layer will not be continuous and source–drain current will not flow, or at least not obey the theoretical predictions). The thin oxide is typically 0.1 μm. The conventional thick oxide, also called the field oxide, is about 1 μm thick and prevents the formation of inversion layers where the gate metal extends to form the connection to other parts of the chip.

8.7 SOME ADVANCED PROCESSING TECHNIQUES

8.7.1 Polysilicon deposition

Polysilicon deposition is widely used in modern technology. If instead of using the conditions for epitaxial growth, which create a crystalline layer on the silicon surface, somewhat different conditions are employed, a layer can be formed that is made up of many small volumes of crystalline silicon, each volume having a different orientation from its neighbor. The shape of the regions is more or less random, as shown in Fig. 8.17.

Polysilicon is normally deposited using low-pressure chemical vapor deposition (LPCVD) at a temperature between 600 and 650°C. The deposition commonly uses silane (SiH_4) with the reaction:

$$SiH_4 \Longrightarrow Si + 2H_2 \tag{8.11}$$

The layers are typically half a micrometer thick and may be doped or undoped.

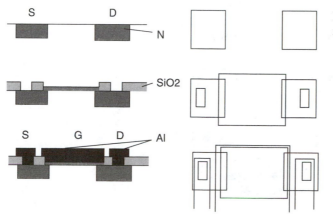

Figure 8.16 Metal gate MOSFET main fabrication steps.

SURFACE

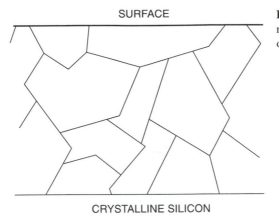

CRYSTALLINE SILICON

Figure 8.17 A polysilicon layer. Inside each region is crystalline silicon with various crystal orientations.

8.7.2 Oxide isolation

Oxide isolation is widely used in advanced technology devices. Figure 8.18 shows the essential steps.

The nitride layer (Si_3N_4) shown in diagram Fig. 8.18(a) is used as a barrier against SiO_2 growth; about 1500 Å is adequate. The bottom SiO_2 layer is to stop the nitride etch. In Fig. 8.18(b), the window has been opened in the area where the oxide isolation is required. In Fig. 8.18(c) the isolation region is etched to a depth equal to 55% of the total oxide depth required. This allows for the fact that the SiO_2 formation consumes silicon. The final oxidation shown in Fig. 8.18(d) is such that the surface is nearly plane—a requirement for subsequent processing, particularly interconnects, where open circuits can occur if the step for a metal is too high.

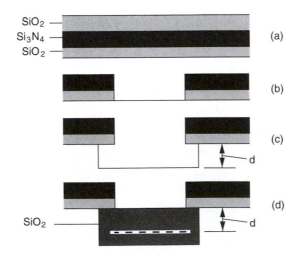

SiO_2 —
Si_3N_4 — (a)
SiO_2 —

(b)

(c) d

(d)

SiO_2 — d

Figure 8.18 Oxide isolation: (a) SiO_2, nitride, SiO_2 layers; (b) after etching window; (c) after etching silicon; (d) after growing SiO_2.

8.7.3 Trench isolation

A more recent form of dielectric isolation uses a vertical trench, created using reactive ion etch (RIE) techniques. This type of etching uses an ion beam to react with the silicon and etch it in a plane perpendicular to the surface. Several variations are used to fill the trench, and Fig. 8.19 shows the general principle.

   An oxide layer is usually grown around the edges of the trench. It is then filled by depositing undoped polysilicon.

8.8 ADVANCED TECHNOLOGY BIPOLAR TRANSISTORS

8.8.1 Polysilicon emitter BJT structures

Most integrated circuit BJT structures are now made using polysilicon for the emitter contact. This allows for self-alignment of the emitter diffusion and emitter "contact" and therefore gives reduced areas, with a corresponding reduction in junction capacitance and hence improved f_t. Figure 8.20 illustrates the essential steps.

   Figure 8.20(a) shows the base already formed, with a window opened in the SiO_2. In Fig. 8.20(b) undoped polysilicon has been deposited over the whole wafer. In Fig. 8.20(c) the polysilicon has been doped by implanting and diffusing phosphorus or arsenic impurities. These diffuse very rapidly along the grain boundaries and only a very shallow junction depth, of order 0.05 μm, is required below the silicon surface. In Fig. 8.20(e), the polysilicon has been removed by a lithographic and etch step, and metal has been deposited and defined as in a conventional process.

8.8.2 Oxide and trench isolation for BJTs

Oxide isolation is used BJT technology to eliminate the sidewall capacitances of diffused or implanted layers.

   The main application of trench isolation in bipolar technology is as a substitute for the deep P^+ isolation diffusion. Trench isolation offers the advantage of reduced capacitance (between

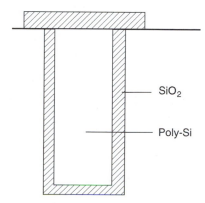

Figure 8.19 Trench isolation, with an SiO_2 layer filled with polysilicon.

SiO$_2$

Poly-Si

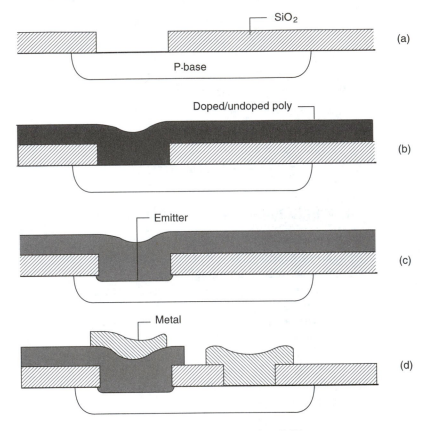

Figure 8.20 Basic steps in the formation of a polysilicon emitter BJT.

collector and substrate) and takes up less area (because of the vertical etch properties of RIE). A typical trench may be 2 μm wide by 6 μm deep. The actual depth must be such that the trench extends down into the P^- substrate.

Figure 8.21 shows an example of a complete BJT structure using advanced technology. The emitter and base regions are both self-aligned using a combination of RIE and spacer oxide to determine the separation. Note the use of trench isolation through the buried layer to reduce collector substrate capacitance. Oxide isolation is used in the active device areas. It can be seen that the actual transistor emitter area is a small fraction of the total area.

8.9 ADVANCED TECHNOLOGY MOSFET STRUCTURES

8.9.1 Polysilicon MOSFET devices

This represents the second half of silicon integrated circuit technology. Today, nearly all MOSFETs are made using polysilicon for the gate, since this offers a self-aligning feature

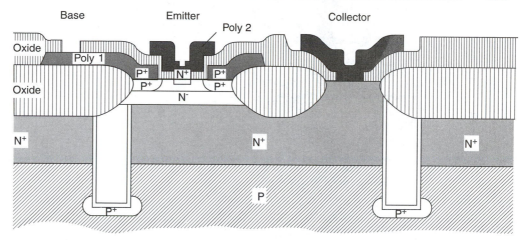

Figure 8.21 Cross-section diagram of an advanced double self-aligned polysilicon emitter BJT with oxide and trench isolation. After M. C. Wilson, P. C. Hunt, S. Duncan, and D. J. Bazley, "10.7 GHz frequency divider using double layer polysilicon process technology," *Electronics Letters*, **24**, 920–22, July, 1988, reproduced with permission from IEE.

similar to that for the BJT emitter described in the preceding section. Figure 8.22 shows the essential steps in this "silicon gate" MOSFET process.

The starting point for an *N*-channel MOSFET circuit is a lightly doped *P* substrate; this will typically be 200 to 400 μm thick. After growing an oxide layer (of order 1 μm thick) a window is opened as shown in Fig. 8.22(a). The surface is carefully prepared and cleaned before growing the critical thin gate oxide, Fig. 8.22(b); in modern devices this is less than 0.1 μm thick, and both its thickness and the surface cleanliness between the *P* silicon and the thin oxide are crucial in determining the electrical characteristics of the MOSFET.

An undoped polysilicon layer is now deposited over the whole wafer, to a thickness of order 0.5 μm as shown in Fig. 8.22(c). A second mask is now used to define the critical gate length (1 μm or less in modern devices), within the thin oxide region. The polysilicon is removed by etching in all areas except the gate. The residual thin oxide is also removed, leaving the structure shown in Fig. 8.22(d), without the source and drain implants.

By introducing *N*-type dopant (either implanted or diffused, phosphorus or arsenic), the polysilicon is doped (to create a low-resistivity gate) and the source and drain junctions are formed by diffusion. Note that it is the edge of the polysilicon region that defines the edge of the source and drain diffusions—in other words, we have a self-aligned structure in which the gate starts exactly at the right-hand side of the source and ends exactly at the left-hand side of the drain, as shown in Fig. 8.22(d). A final oxide, mask, lithographic step is used to define the contact windows, as shown in Fig. 8.22(e); this is followed by a conventional metallization and interconnect step. Note that in Fig. 8.22(e), a metal contact opening is not necessary on the polysilicon over the gate. The doped polysilicon may be used as an interconnect and the contact made at any convenient position.

The *N*-channel MOSFET is now complete. As discussed in Chapter 5, the threshold voltage is affected by the surface charge density between the thin oxide and the *P* silicon

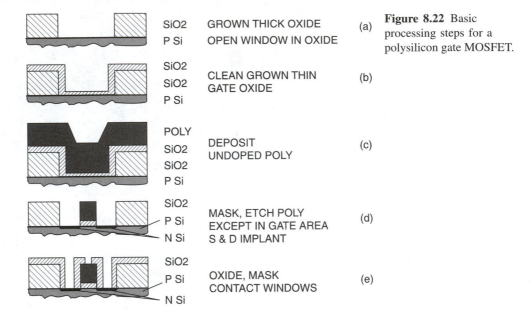

SiO2	GROWN THICK OXIDE (a)
P Si	OPEN WINDOW IN OXIDE

SiO2	CLEAN GROWN THIN (b)
SiO2	GATE OXIDE
P Si	

POLY	DEPOSIT (c)
SiO2	UNDOPED POLY
SiO2	
P Si	

SiO2	MASK, ETCH POLY
P Si	EXCEPT IN GATE AREA (d)
N Si	S & D IMPLANT

SiO2	OXIDE, MASK
P Si	CONTACT WINDOWS (e)
N Si	

Figure 8.22 Basic processing steps for a polysilicon gate MOSFET.

and by the depletion-layer charge, which is a function of doping under the gate oxide. It is therefore possible to adjust the threshold voltage by suitably implanting the channel region prior to depositing the polysilicon.

Since the MOS action occurs only in the channel under the gate, additional MOSFETs can be placed adjacent to each other on a chip with no need for isolation and a consequent high packing density (large number of transistors in a given chip area). This is one important advantage of an NMOS chip over a standard bipolar chip. The number of processing steps is also reduced in NMOS compared to standard bipolar. In the early 1970s NMOS was a very popular technology for these reasons. Speed and power consumption were not, however, good, and CMOS evolved rapidly.

8.9.2 CMOS technology

Complementary MOSFET, or CMOS, technology is extensively used today in digital applications. As mentioned in Chapter 1, a CMOS inverter consists of both an N-channel and a P-channel device, with a common input to the two gates. In order to fabricate this combination, it is necessary to have both a P and an N region available for the channel. This can be obtained in the above N-channel MOSFET technology by first implanting and/or diffusing an "N tub" in the regions where P-channel devices are required. Figure 8.23 shows the cross section of a CMOS structure.

This is apparently not too complicated. However in reality, several more processing steps are essential to obtain good-performance CMOS structures. First, the threshold voltages of the N- and P-channel devices have to be matched for a good dc transfer characteristic. This necessitates the use of at least one extra ion implant step.

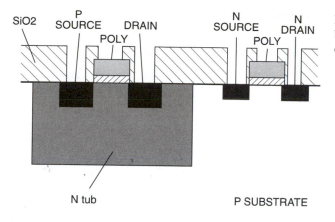

Figure 8.23 Cross section of a basic CMOS structure with an N tub on a P substrate.

The second problem is related to the presence of several P- and N-diffused regions. As explained in Chapter 5, there is a lateral $PNPN$ path available, which can lead to SCR or thyristor action. Latchup occurs when the sum of the two common-base current gains of the PNP and NPN BJTs reaches unity. This can be prevented in two ways: First, if a deep-junction heavy-doped implant or diffusion is made in the appropriate lateral BJT base regions, the gains can be forced never to reach the latch-up condition. This method was used in early CMOS chips. However, a heavy-doped junction layer causes a large parasitic capacitance and also takes up considerable area because of the large sideways diffusion associated with the required deep junction. A better solution is to use a combination of oxide and trench isolation. A cross section of a typical advanced CMOS structure is shown in Fig. 8.24.

The trench isolation effectively eliminates the lateral NPN and PNP current paths, hence eliminating the possibility of latch-up. Note also the use of a lightly doped substrate, which can be either P or N type, provided the doping level of the tubs is slightly greater than the substrate doping level. Oxide "walls" may be used for the P and N regions to reduce parasitic junction capacitance.

8.9.3 BiCMOS technology

Advanced chips often use a combination of CMOS and bipolar devices. The technology is called BiCMOS and requires some extra steps. Figure 8.25 illustrates a possible scenario for combining bipolar and CMOS on one chip. In Fig. 8.25(b) an extra P^+ base diffusion is used;

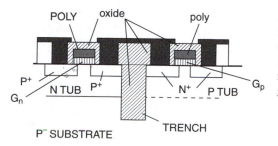

Figure 8.24 Cross section of the essential features of an advanced CMOS structure using oxide and trench isolation (see, for example, [11]). G_n and G_p are the two thin gate oxide regions. Note that the doped polysilicon of the gates is used as an interconnect.

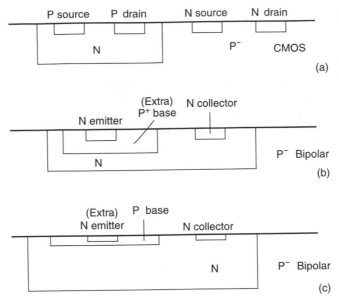

Figure 8.25 A simple BiCMOS combination: (a) standard CMOS; (b) addition of a P^+ base diffusion to create BiCMOS; and (c) alternative implementation with addition of an N^+ emitter diffusion.

in Fig. 8.25(c) an extra N emitter step is added. Many variations are possible for combining the two types of device, but invariably one or more extra processing steps are involved. A modern BiCMOS chip is extremely complex and will have about twenty mask steps, with a minimum feature size (1996) of about 0.5 μm.

8.10 CONCLUSIONS

In this chapter we have provided the reader with an introduction to the more important aspects of processing and fabrication of a silicon chip. The material has been selected to provide the sort of information that (1) assists in understanding the design and operation of the chip, (2) provides the minimum background necessary for the electronics engineer to have dialogue with his chemical engineer counterpart in discussing processing.

We outlined the steps from crystal growth and epitaxial layers to oxide growth and lithography. Diffusion and implantation of impurity atoms was treated in some more detail since the electronics engineer frequently is involved in the prediction of electrical performance as affected by impurity atom distributions. In most cases CAD software such as SUPREM3 [6] is used for this purpose. Examples were given of various integrated circuit chips, from conventional metal-contacted BJT and metal gate MOSFETs to advanced devices using polysilicon deposition, with oxide and trench isolation.

PROBLEMS

1. Calculate the atomic diffusion length and estimate the junction depth for a boron infinite source diffusion into an epitaxial layer doped 10^{15} cm^{-3}. The diffusion is carried out at 1100°C for a duration of 30 min. Assume a solid solubility of 2×10^{20} cm^{-3}.

2. A bipolar transistor made by using the above boron diffusion followed by a limited-source drive-in at 1100°C for 1 h, followed by a phosphorus diffusion at 1100°C for 30 min. Calculate: (a) the total boron concentration integral Q_0 at the end of the predeposition step; (b) 'the surface boron concentration $C_{s'}$ at the end of the drive-in phase; (c) estimate the emitter–base and base–collector junction depths. Make suitable approximations where necessary.

3. Estimate the sheet resistance of the boron and phosphorus diffusions in the above questions. Make suitable approximations.

4. Assume that the sideways phosphorus diffusion in Problem 2 is characterized by a diffusion length 0.8 times the vertical value. Calculate the horizontal value of the emitter junction depth at the surface and hence sketch the metallurgical junction contour in the sidewall region.

5. For a background doping level in 10^{15} cm^{-3} N type, use the information available in Table 8.3 to determine the impurity distribution and to estimate the junction depth for a boron implant with an implant dose of 10^{14} cm^{-2} at an energy of: (a) 300 keV; (b) 30 keV.

References

1. S. K. Ghandhi, *VLSI Fabrication Principles*. New York: John Wiley, 1983.
2. S. M. Sze (Ed.), *VLSI Technology*, 2nd ed. New York: McGraw–Hill, 1988.
3. P. E. Gise and R. Blanchard, *Modern Semiconductor Fabrication Technology*. Englewood Cliffs, NJ: Prentice-Hall, 1986.
4. M. S. Tyagi, *Introduction to Semiconductor Materials and Devices*. New York: John Wiley, 1991.
5 S. M. Sze, *Semiconductor Devices: Physics and Technology*. New York: John Wiley, 1985.
6. D. A. Antoniadis, S. E. Hansen and R. W. Dutton, "SUPREM-II. A program for IC process modeling and simulation," Stanford Electronics Laboratories Technical Report No. 5019–2 (June, 1978).
7. D. P. Kennedy and R. R. O'Brien, "Analysis of the impurity atom distribution near the diffusion mask for a planar *P–N* junction," *IBM Jnl. Res. Dev.* **9**, 179 (1965).
8. J. D. Meindl et al. "Silicon epitaxy and oxidation," in F. Van de Wiele, W. L. Engl and P. O. Jespers, Eds., *Process and Device Modelling for Integrated Circit Design*. Leyden: Noorhoff, 1977.
9. L. Van den Hove, "Optical lithography, how far will it go?" *Proceedings of the 27th European Solid-State Device Research Conference*, Stuttgart, Sept. 1997, Ed. H. Grunbacher, Editions Frontieres, Paris, 1997.
10. K. A. Pickar, "Ion implantation in silicon," in R. Wolfe, Ed., *Applied Solid State Science* **5**, New York: Academic, 1975.
11. R. D. Rung, H. Momose, Y. Nagakubo, "Deep trench isolated cmos devices," *IEEE Electron Device Meeting, Tech. Digest*, p. 237, 1982.

Appendix A

The Bohr Model of the Hydrogen Atom

Consider the Bohr model of the hydrogen atom. The orbiting electron has associated with it a centripetal force

$$F_c = mv^2/r \tag{A.1}$$

counteracted by a coulomb force

$$F_q = \frac{1}{4\pi\epsilon_0} \frac{q^2}{r^2}, \qquad q^2 = q_1 q_2 \tag{A.2}$$

Hence

$$\frac{mv^2}{r} = \frac{1}{4\pi\epsilon_0} \frac{q^2}{r^2} \tag{A.3}$$

$$v = \frac{q}{\sqrt{4\pi\epsilon_0 mr}} \tag{A.4}$$

But the kinetic energy is

$$KE = \tfrac{1}{2}mv^2 \tag{A.5}$$

and the potential energy (force on electron in negative r direction) is:

$$PE = \frac{q^2}{\sqrt{4\pi\epsilon_0 r}} \tag{A.6}$$

The total energy is thus:

$$E = \frac{mv^2}{2} - \frac{q^2}{4\pi\epsilon_0 r} = -\frac{q^2}{8\pi\epsilon_0 r} \tag{A.7}$$

The de Broglie wavelength is

$$\lambda = h/mv \tag{A.8}$$

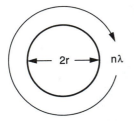

Figure A.1 The Bohr orbit for the hydrogen atom.

and it is reasonable to suppose that for a stable orbit there must be an integral number of de Broglie wavelengths as shown in Fig. A.1, that is,

$$n\lambda = 2\pi r \tag{A.9}$$

We thus obtain

$$nh/mv = 2\pi r \tag{A.10}$$

Substituting for the velocity obtained above, we have

$$2\pi r = \frac{nh}{mq}\sqrt{4\pi\epsilon_o mr} \tag{A.11}$$

that is, the radius r is

$$r_n = \frac{\epsilon_0 h^2}{\pi m q^2}n^2 = 0.53n^2 \text{ Å} \tag{A.12}$$

and the energy E is given by:

$$E_n = \frac{-q^4 m}{8\epsilon_0^2 h^2}\frac{1}{n^2} = \frac{-13.6}{n^2} \text{ eV} \tag{A.13}$$

The electron is thus seen to lie in one of an infinite number of possible discrete orbits, each one corresponding to a defined energy level.

Appendix B

The Schrödinger Wave Equation

In 1926 Schrödinger produced his wave equation of an electron, with potential energy V:

$$\frac{h^2}{8\pi^2 m_0}\left[\frac{\partial^2\Psi}{\partial x^2} + \frac{\partial^2\Psi}{\partial y^2} + \frac{\partial^2\Psi}{\partial z^2}\right] - V\Psi = \frac{h}{2\pi j}\frac{\partial\Psi}{\partial t} \tag{B.1}$$

Here h is the Planck constant, m_0 the electron mass, $j = \sqrt{-1}$, and Ψ is the wave function, which depends in general on position and time; that is,

$$V = V(x, y, z, t) \tag{B.2}$$

$$\Psi = \Psi(x, y, z, t) \tag{B.3}$$

This equation was postulated by Schrödinger in the same way that Newton "postulated" his laws of motion. It cannot be proved, other than heuristically, but fits all experimentally observed situations. Born (1926) showed that $|\Psi|^2$ gives the probability of finding the electron within a distance x to $x + \Delta x$. We will illustrate this for the case of the hydrogen atom (Fig. B.1). The time-independent solution to the Schrödinger wave equation in this case can be shown to be, in terms of the distance r from the center of the atom, is:

$$V = -q^2/4\pi\epsilon r = \text{ potential energy} \tag{B.4}$$

$$\Psi = \pi^{-1/2}\rho^{-3/2}e^{-r/\rho} \tag{B.5}$$

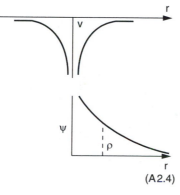

(A2.4)

Figure B.1 Solutions to the wave equation for a single electron orbiting the hydrogen atom.

$$\rho = \frac{h^2 \epsilon_0}{\pi m q^2} = 0.53 \text{ Å} \tag{B.6}$$

This is precisely the value of the Bohr radius.

The probability that the electron lies between a radius r and $r + dr$ is

$$p = |\psi|^2 \, dv \tag{B.7}$$

Since the volume of a sphere is

$$v = \tfrac{4}{3}\pi r^3 \tag{B.8}$$

It follows that

$$dv = 4\pi r^2 \, dr \tag{B.9}$$

and hence

$$P(r) = 4\pi \, |\Psi|^2 \, r^2 \, dr = 4r^2 \rho^{-3} \rho^{-2r/\rho} \, dr \tag{B.10}$$

This is shown in Fig. B.2. Note that now we have the information that the probability peaks in the vicinity of the original Bohr radius. We must henceforth think therefore of the "orbital" as being the "most likely" position for the electron, but in fact it can be virtually "anywhere"; it is just extremely unlikely ($|\psi|^2 r^2 \to 0$) to be near $r = 0$ or near $r = \infty$.

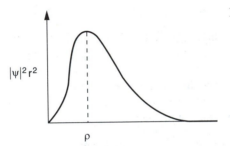

Figure B.2 Plot of probability distribution (B.10).

Appendix C

Electron in a Square Potential Well with Infinite Sides

Consider the electron in a potential well as shown in Fig. C.1. This is a very rough approximation to the electron in an atom, bound within $\pm x_0$. The wave equation is:

$$\frac{d^2\psi}{dx^2} + \frac{8\pi^2 mE}{h^2}\psi = 0 \tag{C.1}$$

In this case, the electron cannot escape; that is, the wave function $\psi = 0$ for $|x| > x_0$, and also $\psi = 0$ at $x = \pm x_0$. For $-x_0 < x < +x_0$, the energy is constant, hence set $P = 0$ (free choice of reference for potential). The solution to Schrödinger's wave equation is thus:

$$\psi = A \sin \frac{2\pi}{h}\sqrt{2mE}x + B \cos \frac{2\pi}{h}\sqrt{2mE}x \tag{C.2}$$

At $x = x_0$,

$$0 = A \sin \beta x_0 + B \cos \beta x_0, \quad \beta = \frac{2\pi}{h}\sqrt{2mE} \tag{C.3}$$

At $x = -x_0$,

$$0 = -A \sin \beta x_0 + B \cos \beta x_0 \tag{C.4}$$

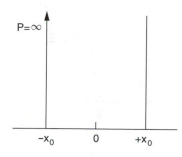

Figure C.1 Infinite potential well.

Hence we find:

$$2B \cos \beta x_0 = 0 \tag{C.5}$$

$$2A \sin \beta x_0 = 0 \tag{C.6}$$

Hence:

$$B = 0 \text{ and } \sin \beta x_0 = 0; \ \beta x_0 = n\pi/2, \qquad n \text{ even} \tag{C.7}$$

or

$$A = 0 \text{ and } \cos \beta x_0 = 0; \ \beta x_0 = n\pi/2, \qquad n \text{ odd} \tag{C.8}$$

The solutions for the wave function are thus:

$$\psi = \begin{cases} A \sin \beta x, & n \text{ even} \\ B \cos \beta x & n \text{ odd} \end{cases} \tag{C.9} \tag{C.10}$$

$$\beta = \frac{n\pi}{2x_0} = \frac{2\pi}{h}\sqrt{2mE} \tag{C.11}$$

for normalization,

$$\int_{-\infty}^{\infty} |\psi^2| \, dx = 1 \tag{C.12}$$

$$E_n = \frac{n^2 h^2}{32 m x_0^2} J \tag{C.13}$$

This leads to

$$A = 1/\sqrt{x_0} = B \tag{C.14}$$

$$\psi_n = \left(1/\sqrt{x_0}\right) \cos \beta x, \qquad n = 1, 3, 4 \tag{C.15}$$

$$\psi_n = \left(1/\sqrt{x_0}\right) \sin \beta x, \qquad n = 2, 4, 6 \tag{C.16}$$

The $n = 1$ and 2 solutions are plotted in Fig. C.2. This results will be used in explaining the concept of energy-level splitting in Appendix D.

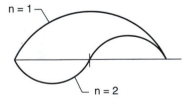

Figure C.2 $n = 1$ and 2 solutions for the infinite potential well.

Appendix D

Splitting of Energy Levels

Based on the preceding square potential well solution to Schrödinger's wave equation, let us now consider qualitatively what happens to the wave functions of two separate "wells" as they are brought together, bearing in mind that this is a (very) crude approximation to two electrons bound to each of two atoms. In order to better approximate the real situation, we consider potential wells with *finite* sides, so that there is a finite probability of the wave function being nonzero outside the wells. It can be shown (quite simply) that outside the wells, providing the electron energy is less than the potential barrier of the wells, an exponential decay occurs. We will consider three cases: (1) far spacing, no interaction, (2) moderate spacing, some interaction, (3) final interatomic spacing. In each case, both positive and negative solutions exist for Ψ (Fig. D.1).

Note that in case (3), the wave functions are identical in appearance to those obtained for the single square well for $n = 1$ and 2, provided the appropriate energy levels are used (Fig. D.2).

$$E_n = (n^2/x_0^2)(h^2/32m)$$

1. Symmetric: $n = 1$, $X = 4x_0 E_1' = E_1 x(1^2)/2^2 = E_1/4$.
2. Antisymmetric: $n = 2$, $X = 4x_0 E_2' = E_1 x(2^2)/2^2 = E_1$.

In the real case of the H_2^+ ion (one electron), with the potential "well" created by the potential energy (inversely proportional to the radial distance) already considered in discussing the H atom, the wave functions appear as in Fig. D.3.

In the case of a solid, the multiple energy levels of individual atoms split into bands of energies, alternating between full and empty bands, as assumed in the discussions of Chapter 2.

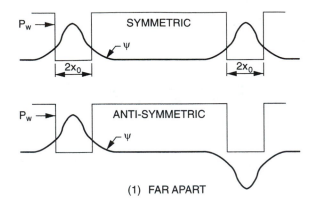

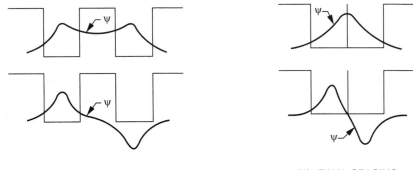

Figure D.1 Symmetric and antisymmetric wave functions for various spacings.

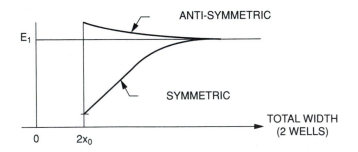

Figure D.2 Actual energy levels versus spacing.

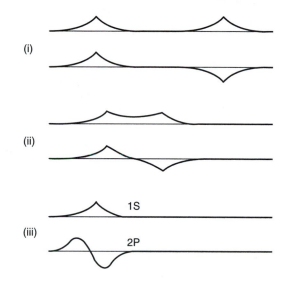

Figure D.3 Actual wave function solution versus spacing.

Appendix E

The Fermi Probability Distribution

In a situation where particles are noninteracting, such as a gas, the Boltzmann probability distribution versus energy is obeyed. In this case:

$$P(E) \propto \exp(-E/kT) \tag{E.1}$$

where k is Boltzmann's constant, E is the energy. Electrons, on the other hand, are constrained by the Pauli exclusion principle (no two electrons can occupy the same quantum state) and are said to be interacting particles. In this case the Fermi distribution must be used to calculate the probability of occupancy of a quantum state by an electron at a given energy. The Fermi function is given by:

$$F(E) = 1/\{1 + \exp[(E - E_F)/kT]\} \tag{E.2}$$

where E_F is the Fermi level.

This distribution is shown below for a moderately doped N-type semiconductor. E_F is the energy level at which there is a 50% probability of occupancy of an available state by an electron. At higher energies the probability tends to zero. At low energies, the probability tends to unity; that is, in the valence band there is a high probability of occupancy; that is, nearly all states are filled. Note that for energies greater than a few kT (say, greater than 0.1 eV) above E_F, the Fermi distribution is approximated very well by the Boltzmann distribution.

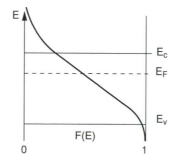

Figure E.1 Fermi probability distribution versus energy for moderate N-type doping.

Appendix F

Density of States and Free-Electron Distribution Versus Energy

The kinetic energy (KE) of an electron in the conduction band $E - E_c$ (see Fig. F.1) is given as a function of the momentum as follows:

$$\tfrac{1}{2}m_e^* v^2 = \frac{p^2}{2m_e^*} = E - E_c \tag{F.1}$$

Here m_e^* is the effective mass of the electron. In quantum mechanical terms, $1/m_e^* = (2\pi/h)^2 \, d^2E/dk^2$, where $k = 2\pi p/h$. The momentum p is given by $p = m_e^* v$; that is,

$$p = \sqrt{2m_e^*(E - E_c)} \tag{F.2}$$

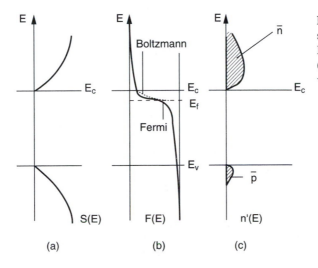

(a) (b) (c)

Figure F.1 (a) Density of available states versus energy; (b) Fermi and Boltzmann probability distributions, (c) electron and hole concentrations versus energy.

In three-dimensional momentum coordinates, a constant kinetic energy corresponds to a spherical surface centered at the origin. The radius of the sphere is p, the magnitude of the momentum. One can then define a differential momentum volume dV_p:

$$dV_p = 4\pi p^2\, dp = 2\pi p\, d(p^2) \tag{F.3}$$

$$= 2\pi \sqrt{2m_e^*(E - E_c)}\, d[2m_e(E - E_c)]$$

$$= 4\sqrt{2}\pi m_e^{*3/2}\sqrt{(E - E_c)}\, dE \tag{F.4}$$

Minimum momentum volume

Heisenberg's uncertainty principle enables us to define the minimum value of momentum for given physical dimensions of the material

$$\Delta P_x = \frac{h}{L_x}$$

$$\Delta P_y = \frac{h}{L_y}$$

$$\Delta P_z = \frac{h}{L_z}$$

where L_x, L_y, and L_z are physical dimensions of a "cube"; ΔP_x, ΔP_y, and ΔP_z are the momentum components in each direction.

$$\Delta P_x \Delta P_y \Delta P_z = \Delta V_p = \frac{h^3}{L_x L_y L_z} = \frac{h^3}{V} \tag{F.5}$$

where V is the volume of the material and ΔV_p is the smallest possible momentum volume that corresponds to one energy level or state. The number of energy levels is therefore given by

$$\frac{d(energy\ levels)}{dV_p} = \frac{V}{h^3} \tag{F.6}$$

Since from Pauli's exclusion principle there are two states per energy level, we can write:

$$d\left(\frac{states}{cm^3}\right) = \frac{2\,dV_P}{h^3}$$

$$= 8\sqrt{2}\pi \frac{m_e^{*3/2}}{h^3}\sqrt{E - E_c}\, dE \tag{F.7}$$

This is the density of states per unit energy shown in Fig. F.1(a). The carrier concentration per unity energy, n', is now given by

$$n'\, dE = F(E)\, d\left(\frac{states}{cm^3}\right) \tag{F.8}$$

where $F(E)$ is the Fermi factor. This can be approximated by the Boltzmann factor in the conduction band for nondegenerate material. It is the probability of occupancy of a state.

The total carrier concentration is obtained by integrating over all energy levels above the conduction-band edge E_c.

$$\bar{n} = \int_{E_c}^{\infty} n' \, dE \tag{F.9}$$

$$\bar{n} = \frac{8\sqrt{2}\pi m_e^{*3/2}}{h^3} \sqrt{kT} e^{(E_F - E_c)/kT} \int_{E_c}^{\infty} \sqrt{\frac{E - E_c}{kT}} e^{-(E - E_c)/kT} \, dE \tag{F.10}$$

We substitute $x = (E - E_c)/kT$ to give:

$$dx = \frac{1}{kT} \, dE \qquad \text{and} \qquad \int_0^{\infty} \sqrt{x} e^{-x} \, dx = \frac{\pi}{2} \tag{F.11}$$

Thus we obtain

$$n = \frac{4\sqrt{2}}{h^3} \left(\pi m_e^* kT\right)^{3/2} e^{(E_F - E_c)/kT} \tag{F.12}$$

A similar analysis applies for holes:

$$\bar{p} = \frac{4\sqrt{2}}{h^3} \left(\pi m_h^* kT\right)^{3/2} e^{(E_F - E_c)/kT} \tag{F.13}$$

The two relationships are normaly written as follows:

$$\bar{n} = N_c e^{(E_c - E_F)/kT} \tag{F.14}$$

$$\bar{p} = N_v e^{(E_v - E_F)/kT} \tag{F.15}$$

N_c, N_v are constants (called the effective density of states) for a given material at a given temperature.

Fermi Level versus doping concentration

From the above results (F.14), (F.15) taking the ratio n/p gives:

$$n/p = \exp[(E_F - E_c - E_V + E_F)/kT] \tag{F.16}$$

hence:

$$2E_F - (E_c + E_V) = kT \ln(n/p) \tag{F.17}$$

or

$$E_F = (E_c + E_V)/2 + (kT/2) \ln(n/p) \tag{F.18}$$

For intrinsic material,

$$n = p \qquad \text{and} \qquad E_F = (E_c + E_V)/2 = E_i \tag{F.19}$$

For doped N-type material using $n = N_D$, $p = n_i^2/N_D$; hence $n/p = (N_D/n_i)^2$. Hence:

$$E_F = (E_c + E_v)/2 + kT \ln(N_D/n_i) \tag{F.20a}$$

For *P*-type material by analogy we find:

$$E_F = (E_c + E_v)/2 - kT \ln(N_A/n_i) \tag{F.20b}$$

Figure F.2 shows how the free-carrier concentration versus energy varies between intrinsic and moderately doped *N*-type material.

For $N_D = 10^{19}$ cm^{-3} in silicon, $E_F = (E_c + E_v)/2 + 0.025 \times 20.7 = 0.52$ eV and the Fermi level is just at the conduction-band edge. For $N_D > 10^{19}$ cm^{-3}, the Fermi level lies above the conduction-band edge, very much like a metal. Similarly for $N_A > 10^{19}$ cm^{-3} the Fermi level moves into the valence band.

It may be noted that the above equations make use of the approximation that $n = N_D$ and $p = N_A$ in *N*- and *P*-type material, respectively. A rigorous treatment giving the concentration of ionized impurity atoms N_D^+, N_A^+ for an ionization energy E_{di} given by Eq. (2.10) yields the result:

$$N_D^+ = N_D\{1 - 1/[1 + (1/g) \exp(E_{FD}/kT)]\} \tag{F.21a}$$

$$N_A^+ = N_A/[1 + g \exp(E_{FD}/kT)] \tag{F.21b}$$

where E_{FD} is the difference between the Fermi level and the impurity level. In *N*-type material $E_{FD} = E_c - E_{di} - E_F$ and $g = 1/2$. In *P*-type material $E_{FD} = E_v + E_{ai} - E_F$ and $g = 4$, where we use E_{ai} to be the ionization energy of acceptor atoms.

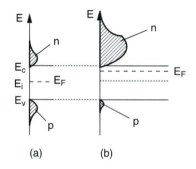

Figure F.2 Electron concentration n and hole concentration p versus energy for: (a) an intrinsic semiconductor; (b) moderately doped *N*-type semiconductor.

Appendix G

The Einstein Relation

Consider a semiconductor bar with nonconstant doping, such as, a *PN* junction, with the *P* side doped N_A, and the *N* side doped N_D. The carrier concentrations will vary in some manner from $p = N_A$ on the *P* side to $p = n_i^2/N_D$ on the *N* side, and similarly for the electrons. Furthermore, the band edge E will change from some value E_{c1} to a value E_{c2}.

Using Eq. (F.14), we can express the ratio of electron concentration at some point x_1 to the concentration at x_2:

$$n = N_c \exp[-(E_c - E_F)/kT] \tag{G.1}$$

$$\therefore n(x_1)/n(x_2) = \exp[-(E_{c1} - E_{c2})/kT] \tag{G.2}$$

Now consider the equation for electron current, drift plus diffusion. In thermal equilibrium, $J_n = 0$, hence:

$$J_n = 0 = q D_n \frac{dn}{dx} + q\mu_n n E \tag{G.3}$$

hence:

$$E = -\frac{D_n}{\mu_n} \frac{1}{n} \frac{dn}{dx} \tag{G.4}$$

Integrating from x_1 to x_2 gives the potential:

$$V(x_1) - V(x_2) = (D_n/\mu_n) \ln[n(x_1)/n(x_2)] \tag{G.5}$$

Comparing this result with the above, and recognizing that energy and potential are related by $E = -qV$, we see that:

$$E_{c1} - E_{c2} = -q[V(x_1) - V(x_2)] \tag{G.6}$$

Therefore,

$$D_n/\mu_n = kT/q \tag{G.7}$$

This is the Einstein relation for electrons. Similarly it may be shown that for holes we have:

$$D_p/\mu_p = kT/q \qquad \text{(G.8)}$$

It is not too surprising that diffusion coefficient and mobility are related, since they are both due to the random thermal motion due to the thermal energy kT of the carriers.

Appendix H

Minority Carrier Diffusion Current and Terminal Current

Consider the narrow-base P^+N diode of Chapter 3. Because the electron current injected into the P^+ region is small, we approximate it to zero. It remains to demonstrate that we can neglect the hole diffusion current on the N side. Setting $J_n = 0$ in Eq. (2.29) gives:

$$E = -V_t \frac{1}{n} \frac{dn}{dx} \tag{H.1}$$

From the discussion on Fig. 3.3 and Eq. (3.3) we can write $n(x) = N_D + p(x)$, and hence the field becomes:

$$E = -V_t/[N_D + p(x)] \frac{dp}{dx} \tag{H.2}$$

The hole drift current is thus:

$$J_{p\text{drift}} = q\mu_p p E = -q D_p \frac{dp}{dx} \frac{p(x)}{N_D + p(x)} \tag{H.3}$$

Define the ratio of hole drift to diffusion current by r_{dd}:

$$r_{dd} = J_{p\text{drift}}/J_{p\text{diff}} = p(x)/[N_D + p(x)] \tag{H.4}$$

Clearly this is much less than unity provided $p(x) \ll N_D$, the condition for low-level injection. We are therefore justified in saying that only minority carrier diffusion current need be considered in the determination of the terminal current of a P^+N narrow base diode.

Appendix I

The Wide-Base Diode Derivation

The equation for excess holes $p'(x)$ diffusing in the presence of recombination is given by Eq. (2.86).

$$\frac{d^2 p'}{dx^2} = \frac{p'(x)}{L_p^2} \tag{I.1}$$

The corresponding diagram for the hole distribution in the P^+N wide-base diode is given in Fig. 3.9. The solution to Eq. (I.1) is:

$$p'(x) = A \exp(-x/L_p) + B \exp(x/L_p) \tag{I.2}$$

Since there is no physical mechanism in this structure creating injection of holes to the right of the N region, it follows that B must zero [otherwise $p'(x)$ would increase with increasing x]. The value of A is the excess hole concentration at $x = 0$ and is obtained from the injection law. Following the procedure used for the narrow-base diode, Eq. (3.2) is the injected hole concentration on the N side of the depletion layer:

$$p_n(0) = p_{n0} \exp(V_a/V_t) \tag{I.3}$$

It follows that

$$p'(x) = [p_n(0) - p_{n0}] \exp(-x/L_p) \tag{I.4}$$

The current flowing at the depletion layer edge, $x = 0$, is the hole diffusion current given by Eq. (2.28):

$$I_p = -q A D_p \left. \frac{dp'}{dx} \right|_{x=0} \tag{I.5}$$

From Eq. (I.4) we obtain:

$$\left. \frac{dp'}{dx} \right|_{x=0} = \frac{-[p_n(0) - p_{n0}]}{L_p} \tag{I.6}$$

The final result for hole current at $x = 0$ is thus:

$$I_p = q A D_p [p_n(0) - p_{n0}] / L_p \tag{I.7}$$

Substituting from (I.3), this may be rewritten in the standard form given by Eqs. (3.23), (3.24).

Appendix J

The *NN⁺* Junction and the *P⁺NN⁺* Diode

Retarding field at the NN⁺ junction

Consider the *NN⁺* junction in thermal equilibrium. The hole distribution for this structure is shown in Fig. 3.10. Since the electron current is zero, Eq. (2.29) gives:

$$0 = qD_n \frac{dn}{dx} + q\mu_n nE \tag{J.1}$$

Rearranging to obtain the electric field and assuming that the region is charge neutral so that $n = N$ (N is a function of x) gives:

$$E = -\left(\frac{D_n}{\mu_n}\right)\left(\frac{1}{N}\right)\frac{dN}{dx} = -\left(\frac{Vt}{N}\right)\frac{dN}{dx} \tag{J.2}$$

where we have substituted $D_n/\mu_n = kT/q = V_t$.

We can now write the hole current density from Eq. (2.30)

$$J_p = -qD_p \frac{dp}{dx} + q\mu_p pE = -qD_p \frac{dp}{dx} - q\mu_p p\left(\frac{V_t}{N}\right)\left(\frac{dN}{dx}\right) \tag{J.3}$$

This can conveniently be rearranged as:

$$J_p = -qD_p \left(\frac{dp}{dx} + \frac{p}{N}\frac{dN}{dx}\right) \tag{J.4}$$

Referring to Fig. 3.10, in the region of the *NN⁺* junction, dp/dx is negative, dN/dx is positive. It is therefore clear that in this equation the diffusion current (the first term) is positive and the drift current (the second term) is negative. The electric field due to the low–high *NN⁺* junction is thus in a direction to oppose the flow of holes. For this reason it is referred to as a "retarding field".

In conclusion, a low–high junction occurring in the direction of minority carrier current flow will introduce a retarding field that opposes current flow and forces the minority carrier concentration down to a low value. Although the above derivation has been done for thermal equilibrium conditions, it is not difficult to accept that the retarding filed will remain unchanged

263

unless high-level injection conditions exist ($n > N$); so the result is generally applicable in most cases.

Minority carrier distribution

We can also argue that the excess hole concentration at the P^+N junction depletion-layer edge on the N side is fixed by the applied bias as given by Eq. (3.2). The holes diffuse across the moderately doped N_{epi} region in the presence of recombination. When the N-type doping level starts to increase rapidly, the hole concentration must decrease at the inverse rate (i.e., $p \propto 1/N_D$). This must be so since in thermal equilibrium, the mass action law $np = n_i^2$ gives:

$$p(x) = n_i^2/N(x) \tag{J.5}$$

As $N(x)$ increases, $p(x)$ decreases. It is not too difficult to show that even in non-thermal equilibrium conditions, $p(x)$ must also decrease. Provided that the distance over which $N(x)$ rises from N_{epi} to N^+ is small compared to the thickness W_n of the moderately doped epitaxial layer, we can assume that $p(x)$ decreases slowly (due to recombination) over the distance $0 < x < W_n$, and decreases rather abruptly for $x > W_n$ (because of the retarding field).

The P+NN+ diode current

The charge Q_p due to excess holes in the region $0 < x < W_n$ can be estimated by assuming only a slight decrease due to recombination, so that $p(W_n)$ (on the left side of the NN^+ junction) is only slightly less than $p(0)$. This condition is satisfied if W_n is not greater than a hole diffusion length L_p and gives:

$$Q_p = qAp'(0)W_n \tag{J.6}$$

The hole concentration on the right-hand side of the $N_{epi}N^+$ interface is very small [from the above discussion); it is approximately $p'(0)N_{epi}/N^+$, which for typical values of $N_{epi} = 10^{15}$ cm^{-3}, $N^+ = 10^{18}$ cm^{-3}, gives $p'(0)/1000$]. It is thus evident that the hole current flowing into the N^+ region must be very small. To a fair approximation for many cases it can be considered as zero. The current created due to the charge Q_p recombining with a hole lifetime τ_p is therefore, from Eq. (2.87):

$$I_p = Q_p/\tau_p \tag{J.7}$$

Substituting for Q_p from Eq. (J.6) and the injection law Eq. (3.2) gives the result contained in Eq. (3.25) and (3.26).

Appendix K

Space-Charge Recombination Current

Referring to Chapter 3, Fig. 3.13, the hole concentration varies over a wide range inside the space-charge layer. Since the Boltzmann relations are quite valid at the low forward bias under consideration, we can use:

$$p(x) = n_i \exp[-V(x)/V_t] \tag{K.1}$$

Furthermore, from the solution to Poisson's equation:

$$V(x) = -(q/\epsilon)N_D(x^2/2 - d_n x) \tag{K.2}$$

For a region close to the junction, $x \ll d_n$ and the potential distribution is given by the approximate form:

$$V(x) = (q/\epsilon)N_D(d_n x) \tag{K.3}$$

It follows that in the vicinity of the junction, the potential varies linearly with distance and therefore that the hole concentration $p(x)$ varies exponentially with distance:

$$p(x) = n_i \exp(-x/x_c) \tag{K.4}$$

where $x_c = (\epsilon V_t)/(q N_D d_n)$

Interestingly, Fig. 3.13 is a fairly good sketch of $p(x)$, at least near $x = 0$. Furthermore, by examining the ratio x_c/d_n, we see that:

$$x_c/d_n = (\epsilon V_t)/(q N_D d_n^2) \tag{K.5}$$

Since the total voltage across the junction $V_{jtot} = V_{bi} - V_a$ and $V_{jtot} = (q/\epsilon)N_D(d_n^2)$, we see that the ratio

$$x_c/d_n = V_t/(V_{bi} - V_a) \tag{K.6}$$

For low forward bias, say, up to around 0.4 V, this ratio is less than 0.1. We can therefore obtain quite an accurate estimate of the total electron charge on the P side:

$$Q_p = \int q A p(x) \, dx \tag{K.7}$$

$$= q A n_c x_c \tag{K.8}$$

The current due to recombination (using a factor of two to account for the other side of the junction) is therefore:

$$I_{\text{scr}} = Q_p/\tau = q A n_c x_c/\tau = (q A n_i x_c/\tau) \exp(V_a/2V_t) \tag{K.9}$$

Appendix L

Maximum Oscillation Frequency of the BJT

Here we give a simplified derivation of the unity power gain, or maximum oscillation frequency, of the BJT. Referring to the hybrid π equivalent circuit of Fig. 4.23, consider the case where the frequency is sufficiently high such that $\omega C_\pi \gg 1/r_\pi$, where $C_\pi = (C_{\text{diff}} + C_{\text{je}})$ and also $\omega C_\pi \gg 1/r_{\text{bb}}$. The small-signal output resistance r_0 is defined as the resistance looking into the collector terminal due to a current i_o fed in at the output node and is obtained by writing:

$$i_0 = g_{\text{m}} v'_{\text{be}} + v_0 j \omega C_{\text{jc}} \tag{L.1}$$

$$i_0/v_0 = g_{\text{m}} v'_{\text{be}}/v_0 + j \omega C_{\text{jc}} \tag{L.2}$$

The real component of this is the output conductance g_0:

$$g_0 = g_{\text{m}} v'_{\text{be}}/v_0 \tag{L.3}$$

The ratio v'_{be}/v_0 may be obtained by considering the potential divider created by C_{jc} in series with C_π. The voltage v_0 across the output node due to the current i_0 creates a voltage v'_{be} at the C_π node where:

$$v'_{\text{be}} = v_0 C_{\text{jc}}/C_\pi \tag{L.4}$$

The output resistance is thus:

$$r_0 = 1/g_0 = C_\pi/(C_{\text{jc}} g_{\text{m}}) \tag{L.5}$$

For maximum power gain G_p, the load resistance must be equal to this output resistance, and the output power is then $i_c^2 r_0/4$. At high frequencies, the input resistance is simply r_{bb}, and the input power is $i_b^2 r_{\text{bb}}$. Using the fact that $v'_{\text{be}} = i_b j \omega C_\pi$, the power gain is thus:

$$G_p = g_{\text{m}}^2 r_0/(4 r_{\text{bb}} \omega^2 C_\pi^2) \tag{L.6}$$

Substituting for r_0 gives:

$$G_p = g_{\text{m}}/(4 r_{\text{bb}} \omega^2 C_\pi C_{\text{jc}}) \tag{L.7}$$

Setting $G_p = 1$, solving for the frequency, and substituting $1/2\pi f_{\text{t}} = C_\pi/g_{\text{m}}$ gives:

$$f_{maxosc} = [f_t/(8\pi r_{\text{bb}} C_{\text{jc}})]^{1/2} \tag{L.8}$$

Appendix M

ECL Propagation Delay Time

The general expression for ECL propagation delay time is given in McGregor, Roulston, et al., *Solid State Electronics* **36**, 391–96, March 1993.

$$t_{pd} = r_{bb}(C_{je} + C_{jc}) + 2r_{bb}t_f/R_L + R_L(1.5C_{jc} + C_{sub} + C_L) + \alpha t_f \tag{M.1}$$

where C_{sub} is the collector–substrate capacitance in an integrated circuit BJT and R_L is the circuit load resistance. The optimum R_L value for minimum delay is:

$$R_L = \left[2r_{bb}t_f/(1.5C_{jc} + C_{sub} + C_L)\right]^{1/2} \tag{M.2}$$

The corresponding dc current is:

$$I_{EE} = V_L\left[(1.5C_{jc} + C_{sub} + C_L)/2r_{bb}t_f\right]^{1/2} \tag{M.3}$$

Note that the large signal transconductance

$$G_m = I_{EE}/(V_L/2) = 2/R_L \tag{M.4}$$

The minimum propagation delay time is:

$$t_{pd,min} = r_{bb}(C_{je} + 3C_{jc}) + 2\sqrt{3r_{bb}C_{jc}t_f} + \alpha t_f \tag{M.5}$$

$$= r_{bb}(C_{je} + 3C_{jc}) + \sqrt{3/\omega_{max}} + \alpha t_f \tag{M.6}$$

Note that for a BJT operating near the peak of the f_t versus I_c curve, $f_t \approx 1/\alpha t_f$, and thus we have:

$$t_{pd,min} \approx R_b C_{je} + 0.27/f_{max} + 1/f_t \tag{M.7}$$

Appendix N

Noise in Semiconductor Devices

There are three main sources of noise in semiconductor devices: thermal noise due to random fluctuation of electrons in a resistance, shot noise due to the random rate of arrival of electrons when direct current flows across a space-charge region, and $1/f$ noise due to surface effects.

N.1 THERMAL NOISE

Thermal noise, also known as Johnson noise and Nyquist noise, is generated due to the random voltage generated across any resistive material due to the random movement of electrons with thermal energy kT. Nyquist derived the formula for the resulting noise power P_{nt}:

$$P_{nt} = kT \, df \tag{N.1}$$

where df is the frequency interval of the system in which the noise is observed. In an amplifier, for example, the noise power will be limited by the bandwidth, and so the noise power from a resistance is often written in the form:

$$P_{nt} = kT \, B_n \tag{N.2}$$

where B_n is the noise bandwidth. This is not identical to the 3 dB bandwidth of the amplifier but is generally close. Only in a system with a "rectangular" frequency response with an abrupt cut-off will the noise bandwidth and the signal bandwidth be the same.

In order to calculate the noise performance of a circuit or device, it is necessary to express the thermal noise as an equivalent voltage or current source. Since the noise is a random fluctuation, we use the *mean noise voltage squared*, v_{ns}^2, or the *mean noise current squared* i_{ns}^2. Since the power available from a resistance R with a series voltage v_{nr} is

$$P_{nt} = v_{nr}^2/4R \tag{N.3}$$

we can express the mean noise voltage squared as:

$$v_{nr}^2 = 4kT R B_n \tag{N.4}$$

Similarly, since the power available from a resistance with a shunt current i_{ns} is

$$P_{nt} = i_{nr}^2 R/4 \tag{N.5}$$

the equivalent mean noise current squared as:

$$i_{nr}^2 = 4kT B_n / R \tag{N.6}$$

The noise equivalent circuit of a resistance R may thus be drawn in either the series voltage or shunt current form, as shown in Fig. N.1.

N.2 SHOT NOISE

Shot noise is due to the random rate of arrival of electrons flowing across a space-charge layer to create a direct current I_{dc}. This means that the direct current is actually made up of a large number of narrow pulses, and therefore a noise is superimposed on the dc level. The magnitude of the noise current squared is given by:

$$i_{ns}^2 = 2q I_{dc} B_n \tag{N.7}$$

N.3 1/f NOISE

1/f noise, sometimes called flicker noise, is due mainly to surface effects related to carrier trapping. It is difficult to quantify, and empirical expressions are normally used. It is important only at very low frequencies of order 1 to 100 kHz.

N.4 NOISE IN A *PN* JUNCTION DIODE

The noise equivalent circuit of a *PN* junction diode is shown in Fig. N.2. The values of the shot noise current is given by:

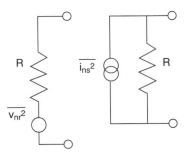

Figure N.1 Series voltage and shunt current noise equivalent circuits of a resistance R.

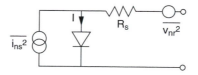

Figure N.2 Noise equivalent circuit of a diode.

$$i_{ns}^2 = 2qIB_n \tag{N.8}$$

The series noise voltage due to the diode series resistance is given by:

$$v_{nr}^2 = 4kTR_sB_n \tag{N.9}$$

Under reverse bias the diode current is the reverse saturation value I_0. The noise current $i_{ns}^2 = 2qI_0B_n$ is often the dominant noise source in photodetectors.

Under forward bias it is interesting to note that if the small-signal equivalent circuit of the diode is used (as in Fig. 3.23), with a small-signal resistance $r_d = V_t/mI$, it is not permissible to use the thermal noise formula (which is for a physical "fixed" resistance). If we did, the thermal shunt noise current squared would be given by:

$$i_{nr}^2 = 4kTB_n/r_d = 4kTB_n/(mV_t/I) = 4kTB_nI/m(kT/q) = 4qIB_n/m \tag{N.10}$$

This is not the same as the (correct) shot noise current given by Eq. (N.8). It does, however, demonstrate the connection between the two sources of noise, both due to random movement of electrons.

N.5 NOISE FIGURE OF A BIPOLAR TRANSISTOR

Figure N.3 shows the noise equivalent circuit of a bipolar transistor with a generator resistance R_g. The voltage v_{ben}^2 is the noise voltage squared across r_π due to the generator noise voltage v_{ng}^2 and the base resistance noise voltage squared v_{nb}^2, together with the voltage due to the shunt shot noise current squared i_{nbs}^2 created by the dc base current I_B.

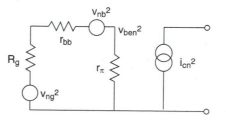

Figure N.3 Noise equivalent circuit for the bipolar transistor.

In amplifiers it is customary to define the noise figure F_n as the ratio of the noise at the output due to all noise sources divided by the noise at the output due only to the generator thermal noise at 300 K.

The value of the noise voltage across r_π is thus given by:

$$v_{\text{ben}}^2 = (v_{\text{nb}}^2 + v_{\text{ng}}^2)\left[r_\pi/(R_\text{g} + r_{\text{bb}} + r_\pi)\right]^2 + i_{\text{nbs}}^2\left[r_\pi(R_\text{g} + r_{\text{bb}})/(r_\pi + R_\text{g} + r_{\text{bb}})\right]^2 \quad \text{(N.11)}$$

The corresponding output (collector) noise current squared is

$$i_{\text{nc}}^2 = g_{\text{m}}^2 v_{\text{ben}}^2 + 2q I_\text{C} B_n \quad \text{(N.12)}$$

where we have added the shot noise due to I_C flowing across the base–collector space-charge layer. Substituting the thermal noise voltages for v_{nb}^2 and v_{ng}^2 and the shot noise current for i_{nbs}^2 and dividing by the noise due only to v_{rg}^2 gives the noise figure:

$$F_n = 1 + r_{\text{bb}}/R_\text{g} + (I_\text{B}/2V_\text{t})(R_\text{g} + r_{\text{bb}})^2/R_\text{g} + (I_\text{C}/2V_\text{t})(R_\text{g} + r_{\text{bb}} + r_\pi)^2/(g_{\text{m}}^2 r_\pi^2 R_\text{g}) \quad \text{(N.13)}$$

where we have substituted V_t for kT/q.

Examination of the first two terms shows that if R_g is large, the contribution from the r_{bb} thermal noise can be decreased. However, increasing R_g will eventually increase the contribution from the base current shot noise since the numerator eventually increases as R_g^2. An optimum value of source resistance thus exists for a given transistor. If we examine the last two terms, we see that if I_C is decreased (to reduce I_B in the second term), the third term (due to the I_C shot noise) will eventually increase due to g_{m}^2 in the denominator ($g_{\text{m}} = I_\text{C}/V_\text{t}$). Optimization of a transistor amplifier thus involves careful design to minimize the noise figure.

Finally it may be noted that the noise figure is usually expressed in dB:

$$F_{\text{ndB}} = 10\ \log_{10} F_n \quad \text{(N.14)}$$

The noise figure may also be expressed in terms of the noise temperature T_n:

$$F_n = 1 + T_n/T_0 \quad \text{(N.15)}$$

Appendix O

The BJT Gummel Integral

This is a general solution for the collector current in a bipolar transistor. We start with the equation for the vertical electron current in the base region of an *NPN* transistor (2.29):

$$J_n = q D_n \frac{dn}{dx} + q \mu_n n E \tag{O.1}$$

In a BJT the hole current is zero in the x direction; so we can write from Eq. (2.30)

$$0 = -q D_p \frac{dp}{dx} + q \mu_p p E \tag{O.2}$$

Rearranging and substituting the Einstein relation (2.31) for D_p/μ_p gives:

$$E = -V_t \left(\frac{1}{p} \right) \frac{dp}{dx} \tag{O.3}$$

This expression for electric field can now be used in Eq. (O.1):

$$J_n = q D_n \left[\frac{dn}{dx} + \left(\frac{n}{p} \right) \frac{dp}{dx} \right] \tag{O.4}$$

This can be written as:

$$\frac{J_n p}{q D_n} = p \left(\frac{dn}{dx} \right) + n \left(\frac{dp}{dx} \right) \tag{O.5}$$

The right-hand side is the derivative $d(np)/dx$. Integrating over the base width from 0 to W_b and treating D_n as a constant gives:

$$(J_n/q D_n) \int_0^{W_b} p \, dx = np(x = 0) - np(x = W_b) \tag{O.6}$$

Under low-level injection, $p(x) = N_A(x)$ and the electron current is thus:

$$I_n = -(q A D_n/G_b) \left[np(0) - np(W_b) \right] \tag{O.7}$$

where

$$G_b = \int_0^{W_b} N_A(x) \, dx \tag{O.8}$$

This is the Gummel integral, that is, the integral of the majority carrier (hole) concentration over the neutral base width W_b. At each junction we can use the injection law (2.73) to give:

$$np(0) = n_i^2 \exp(V_{BE}/V_t); \quad np(W_b) = n_i^2 \exp(V_{BC}/V_t) \tag{O.9}$$

Substituting these relations and using the fact that $I_C = -I_n$ gives:

$$I_C = \left(q A D_n n_i^2 / G_b\right) \left[\exp\left(V_{BE}/V_t\right) - \exp\left(V_{BC}/V_t\right)\right] \tag{O.10}$$

This is a more general form of Eqs. (4.30), (4.108), and (4.109), where $G_b = N_{AB} W_b$.

Appendix P

MOS Inversion-Layer Charge as a Function of Surface Potential

In the following treatment we refer to the MOS structure discussed in Chapter 5. The electric field at the surface E_S of the MOS structure is related to the total surface charge Q_S by Gauss's law:

$$Q_S = -\epsilon E_S \tag{P.1}$$

The total charge is made up of the inversion layer charge Q_I and the depletion layer charge Q_B.

The surface field E_S is determined by solving Poisson's equation for the depletion layer. Considering the case of a P substrate and including the charge due to the free electrons, but neglecting the hole charge, we have:.

$$\frac{dE_x}{dx} = (q/\epsilon)(N_A + n)$$
$$= (q/\epsilon)N_A(1 + n/N_A) \tag{P.2}$$

The electron and hole concentrations are given as a function of potential V by the Boltzmann relations:

$$n = n_i \exp[(V - \phi_F)/kT)] \tag{P.3}$$

$$p = n_i \exp[-(V - \phi_F)/kT)] \tag{P.4}$$

(when $V = \phi_F$ the electron concentration $n = n_i$). Taking the reference potential $V = 0$ in the substrate where $p = N_A$ gives:

$$N_A/n_i = \exp(q\phi_F/kT) \tag{P.5}$$

$$n/N_A = \exp[(V - 2\phi_F)/V_t] \tag{P.6}$$

Thus Eq. (P.2) becomes:

$$\frac{dE_x}{dx} = \frac{qN_A}{\epsilon}\left(1 + e^{(V-2\phi_F)/V_t}\right) \tag{P.7}$$

Noting that $E_x = -dV/dx$ and using the fact that

$$\frac{1}{2} \frac{d}{dx} \left(\frac{dV}{dx}\right)^2 = \frac{dV}{dx} \frac{d^2V}{dx^2} \tag{P.8}$$

gives:

$$\frac{1}{2} \frac{d}{dx} \left(\frac{dV}{dx}\right)^2 = \frac{qN_A}{\epsilon} \frac{dV}{dx} \left(1 + e^{(V-2\phi_F)/V_t}\right) \tag{P.9}$$

This may now be integrated from the substrate neutral region where $V = 0$ and $dV/dx = 0$ to the surface where $dV/dx = -E_x$ and $V = V_S$ to give:

$$\left(\frac{dV_S}{dx}\right)^2 = \frac{2qN_A}{\epsilon} \left[V_S + V_t \left(e^{V_S/V_t} - 1\right) e^{-2\phi_F/V_t}\right] \tag{P.10}$$

Thus $E_S = -dV_S/dx$ is obtained. Combining the result with Eq. (P.1) gives:

$$Q_S = V_t \left(\frac{2q\epsilon N_A}{V_t}\right)^{1/2} \left[V_S/V_t + \left(e^{V_S/V_t} - 1\right) e^{-2\phi_F/V_t}\right]^{1/2} \tag{P.11}$$

The charge Q_B in the depletion layer is $qN_A d_p$, which from Eq. (5.49) gives:

$$Q_B = -(2\epsilon q N_A V_S)^{1/2} \tag{P.12}$$

Since the total charge $Q_S = Q_I + Q_B$ or $Q_I = Q_S - Q_B$, we obtain:

$$Q_I = V_t \left(\frac{2q\epsilon N_A}{V_t}\right)^{1/2} \left\{\left[V_S/V_t + \left(e^{V_S/V_t} - 1\right) e^{-2\phi_F/V_t}\right]^{1/2} - (V_S/V_t)^{1/2}\right\} \tag{P.13}$$

For a given N_A, ϕ_F is fixed and hence Eq. (P.13) gives Q_I as a function of V_S.
   To relate Q_I and V_S to the actual gate voltage V_G we use Gauss's law again:

$$\epsilon_{0x} E_{0x} = -(Q_I + Q_B) \tag{P.14}$$

where the electric field across the thin oxide is:

$$E_{0x} = -(V_G - V_S - V_{FB})/t_{0x} \tag{P.15}$$

Hence:

$$(V_G - V_S - V_{FB}) = -(Q_I + Q_B)t_{0x}/\epsilon_{0x} \tag{P.16}$$

and therefore

$$V_G = V_S + V_{FB} + Q_S/C_{0x} \tag{P.17}$$

where Q_S as a function of V_S is given by Eq. (P.11)

Appendix Q

Student BIPOLE on CD-ROM

The CD-ROM supplied with this book contains the Student BIPOLE software together with many worked examples of diodes, BJTs and MOSFETs for use on a PC with the BIPGRAPH post-processor.

The README.TXT file contains instructions for copying to the PC. The files may be copied using either Windows Explorer or in DOS by typing: xcopy d:/s where d: is the drive label corresponding to the CD-ROM. Both DOS and Windows options are available. Comprehensive documentation in the form of a User's Manual and "Examples" description are contained in both Word, WordPerfect and HTML format (in the \MANUAL directory of the CD-ROM).

It is recommended that the reader start with the worked examples using BIPGRAPH. There are 8 diode, 2 photodiode, 9 BJT and 11 MOSFET examples which have been pre-executed with Student BIPOLE and also 4 BJT examples executed with the industrial BIPOLE3 software. The results of the simulation for these 36 examples may be examined with DOS or Windows graphics. Graphs include physical model parameters such as mobility and carrier lifetime versus doping level, impurity profile and mask layouts, internal characteristics versus bias and

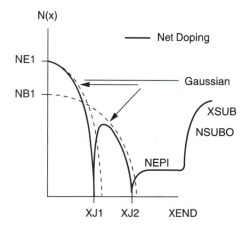

Figure Q.1 Student BIPOLE impurity profile definitions for the BJT.

versus vertical and/or horizontal distance. Finally, a wide range of terminal characteristics are available including both current-voltage and small signal high frequency parameters.

After acquiring familiarity with the various graphs available, the reader may then proceed to alter some of the parameters in the device input files and run Student BIPOLE to generate new data for examination with BIPGRAPH. In this manner it is possible to obtain clear insight into the physical operation of these devices.

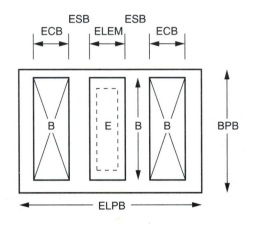

Figure Q.2 Student BIPOLE mask layout definitions for the BJT.

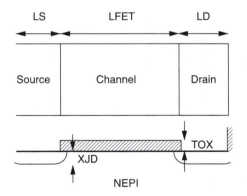

Figure Q.3 Student BIPOLE mask and cross-section definitions for the MOSFET.

Appendix R

Some Useful Physical Constants

Permittivity of free space	ϵ_0	8.85×10^{-14} F/cm
Electron charge	q	1.60×10^{-19} C
Boltzmann's constant	k	1.38×10^{-23} J/K
Thermal voltage at 300 K	V_t	.0259 V
Electron rest mass	m_0	9.11×10^{-31} kg
Planck's constant	h	6.63×10^{-34} J s
Speed of light	c	3×10^8 m/s

Appendix S

List of Important Symbols

C_j junction (depletion-layer) capacitance (F)

C_{diff} diffusion capacitance (F)

D_n, D_p electron, hole diffusion coefficients (cm^2/s)

d_n, d_p depletion-layer thicknesses on n and p sides of a junction (cm)

E_g Band gap of a semiconductor (eV)

E_c, E_v Conduction-, valence-band edges (eV)

E_F Fermi level (eV)

E electric field (V/cm) or energy (eV)

f_t unity current gain frequency (Hz)

g_m transconductance (A/V)

I_0 reverse saturation (leakage) current of a PN junction (A)

J current density (A/cm^2)

L_n, L_p electron, hole diffusion lengths (cm)

m_e^*, m_h^* effective mass of electron, hole (kg)

N_A, N_D Acceptor, donor doping levels (cm^{-3})

N_c, N_v effective density of states (approx 10^{19} cm^{-3})

n, p free-electron, free-hole concentrations (cm^{-3})

n_i intrinsic carrier concentration (in pure material) (cm^{-3})

n_n, n_p free-electron concentrations in n-type, p-type material (cm^{-3})

p_p, p_n free-hole concentrations in p-type, n-type material (cm^{-3})

T temperature (K)

t_{bb} diffusion transit time $= W^2/2D$ (sec)

V_a applied bias (V)

V_A BJT Early voltage

V_{FB} flat-band voltage in MOSFET (V)

V_t thermal volts $= kT/q = 0.0259$ v at 300 K

V_{th} threshold voltage of a MOSFET (V)

V_{bi} built-in barrier potential of a PN junction (V)

V_{jtot} total voltage across a junction $= V_{bi} - V_a$ (V)

v_s saturated drift velocity (cm/s)

v_d drift velocity (cm/s)

v_{th} thermal velocity

W_n, W_i thickness of n region, intrinsic region (of a junction) (cm)

μ_n, μ_p electron, hole mobility (cm^2/V s)

β common-emitter current gain of a BJT

ϕ_F Fermi potential (V)

ϕ_M work function of metal (V)

ϕ_S work function of semiconductor (V)

$\Phi(x)$ photon flux density at position x (photons/cm^2)

x electron affinity (volts)

ν frequency of photons (light) (Hz)

τ carrier lifetime (sec)

ρ resistivity (ohm cm)

σ conductivity (ohm cm)$^{-1}$

Index